Other books you might enjoy:

The Perfect Nonogram Puzzle Book For Adults

All text and puzzles Copyright ©2020 Dave Kinzer

ISBN: 9798565412918

All rights reserved. No part of this book may be reproduced in any form or by any means, without permission from the author.

Welcome to the wonderful world of Nonograms! These puzzles are also called Griddlers, Picross, Hanjie, Japanese Crosswords, Pic-A-Pix, Paint By Numbers, and other names.

If you are new to Nonograms, do the sample puzzle and follow along with my instructions in the next few pages. You'll get the hang of it in no time.

And for a different kind of puzzle challenge, check out my Picture Search Puzzle Book (ISBN: 9798564024853). It's full of puzzles that are similar to word searches, but you look for a series of pictures instead of words.

Try the sample puzzle below!

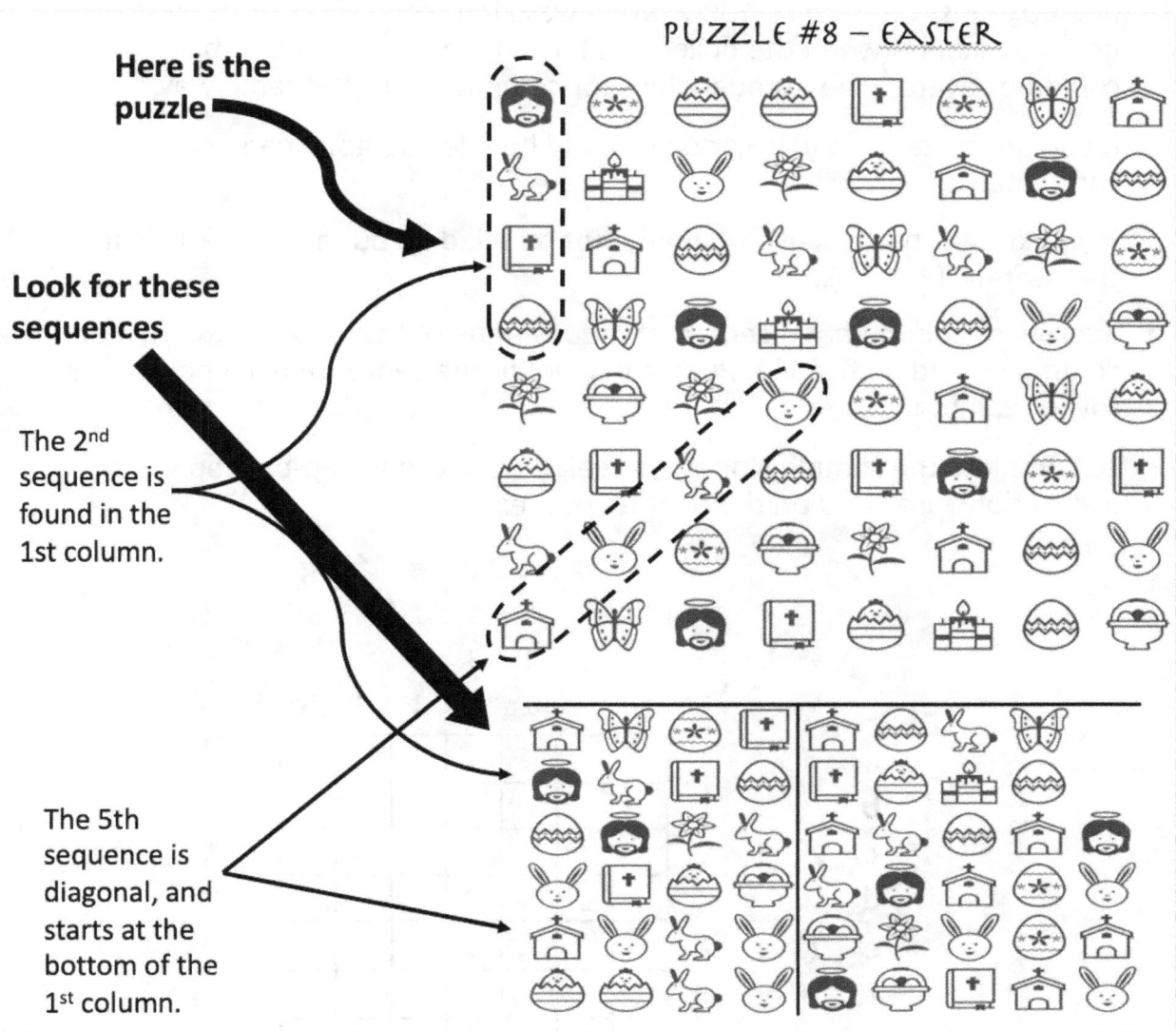

How to complete a Nonogram

- The goal is to fill in the grid, one box at a time, until you have revealed a picture or design. You figure out which boxes to fill in by using logic and the numbers along the outside as clues.

- If you are new to Nonograms, take a look at the basic rules and techniques listed below.

- Each row and column has a number or several numbers in it. The numbers indicates how many boxes in a row must be filled in. For example, if the number clue in a row was "4", it means that at some point in that row, four boxes in a row must be filled in. The rest of the boxes would stay blank. If a row has a clue of 2-3-1, it would mean that somewhere in the row (from left to right), two boxes in a row must be filled in, then you skip <u>at least</u> one box, then three boxes in a row must be filled in, then skip <u>at least</u> one box, then one box must be filled in. You won't always immediately know how many boxes to skip between the boxes you fill in. It could be just one box, or it could be several. The number clues for columns work the same way.

- If you are certain a particular box should be filled in, go ahead and fill it in completely.

- If you know a particular box should **not** be filled in, put a small dot right in the center of the box.

- If a row or column has a zero, that means none of the boxes in the row or column should be filled in. Put a small dot in the center of each box in that row or column.

- To complete the sample Nonogram below, follow my step-by-step instructions provided on the next few pages.

				a	b	c	d	e	f	g
						1	1	1		
						1	3	1		
				5	3	1	1	1	3	5
a			3							
b		1	1							
c	2	1	2							
d			7							
e	2	1	2							
f		1	1							
g			3							

- One of the most helpful techniques to use when starting a Nonogram is to look for the largest numbers in the rows and columns. The largest number used as a clue in our sample Nonogram is a seven. It is in row d. Since this Nonogram is 7x7, this clue in row d tells you that every single box in row d should be filled in. Now your Nonogram should look like this:

				a	b	c	d	e	f	g
						1	1	1		
						1	3	1		
				5	3	1	1	1	3	5
a			3							
b		1	1							
c	2	1	2							
d			7	■	■	■	■	■	■	■
e	2	1	2							
f		1	1							
g			3							

- Now look for other large numbers. You'll see a five for a clue in column a and g. A large number may help you determine some of the boxes to fill in, but not all. You know that column a should have five boxes filled in, you're just not sure *which* five. How many different ways can you fill in five consecutive boxes in column a? There are three different ways, like this:

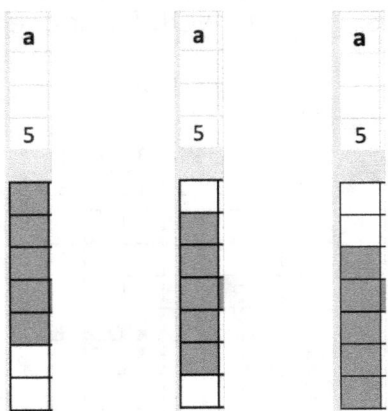

- Those are the only three possibilities. Now look for boxes that were filled in on each example. Notice how the 3rd, 4th, and 5th boxes (from the top) were filled in on each of the three examples? That means that you can be certain that those boxes will be filled in. You're still not sure about the top two and bottom two boxes in column a though. We just leave them blank for now.

- You won't put a dot in those boxes because you're not certain which of them will remain blank. Since column g also has a clue of 5, you know that this logic applies to that column as well. Fill in the 3rd, 4th, and 5th boxes (from the top for both column a and g).

- Now your Nonogram should look like this:

			a	b	c	d	e	f	g
					1	1	1		
				1	3	1			
			5	3	1	1	1	3	5
a		3							
b	1	1							
c	2	1	2	■					■
d		7	■	■	■	■	■	■	■
e	2	1	2	■					■
f	1	1							
g		3							

- Since there aren't any other large numbers in the clues, you'll have to use another strategy. Look at row c. The number clue is 2-1-2. This means there will be a group of two squares filled in, then at least one empty square, then one filled in, then at least one empty square, then two squares filled in. But you can see that the very first square in the row is already filled in. So you know the next square must be filled in as well, because the first number clue in row c is a two. The last two squares in row c must be filled in because the last number clue for row c is a two as well. Fill those in. Now you get this:

			a	b	c	d	e	f	g
					1	1	1		
				1	3	1			
			5	3	1	1	1	3	5
a		3							
b	1	1							
c	2	1	2	■	■				■	■
d		7	■	■	■	■	■	■	■
e	2	1	2	■	■				■	■
f	1	1							
g		3							

- You still need to figure out one more box to fill in for row c, because the clue is 2-1-2. But now you know there is only one box that could be filled in, because there must be at least one empty square between the boxes filled in from the clues. So after the first two filled in boxes for row c, leave an empty box, then fill in the next box, then leave an empty box, and now row c is finished! Put a small dot in the two boxes that you know are empty. The dots will help you figure out columns c and e later. Since row e is identical to row c, fill it out the same way as well.

- Now your Nonogram looks like this:

				a	b	c	d	e	f	g
						1	1	1		
						1	3	1		
				5	3	1	1	1	3	5
a			3							
b		1	1							
c	2	1	2	■	■		.		.	■	■
d			7	■	■	■	■	■	■	■
e	2	1	2	■	■		.		.	■	■
f		1	1							
g			3							

- Finishing rows c and e helped finish two columns as well. Do you see which two columns are done now? It's column b and f, because they need three boxes filled in, and that's already done. Once you add dots to the remaining boxes in those columns that you know shouldn't be filled in, you will know how to fill in the rest of the boxes for two rows. Look closely and see if you can figure out which two rows you can finish. Once you figure it out, see if your Nonogram looks like this:

				a	b	c	d	e	f	g
						1	1	1		
						1	3	1		
				5	3	1	1	1	3	5
a			3	.	.	■	■	■	.	.
b		1	1		.				.	
c	2	1	2	■	■	.	■	.	■	■
d			7	■	■	■	■	■	■	■
e	2	1	2	■	■	.	■	.	■	■
f		1	1		.				.	
g			3	.	.	■	■	■	.	.

- How did you know to fill in rows a and g? Because once you put the dots in the empty boxes in column b and f, the only place you could fill in three boxes in a row was in the middle of the row. When you finish row a and g with dots in the empty boxes, you'll see how to finish column a and g. And *that* actually finished off rows b and f. And just like that, the Nonogram is finished! Not so hard, is it? Erase your dots, and you get this geometric design:

				a	b	c	d	e	f	g
						1	1	1		
						1	3	1		
				5	3	1	1	1	3	5
a			3			■	■	■		
b		1	1	■						■
c	2	1	2	■	■		■		■	■
d			7	■	■	■	■	■	■	■
e	2	1	2	■	■		■		■	■
f		1	1	■						■
g			3			■	■	■		

- Now you can use these same techniques to solve the rest of the puzzles. Have fun!

#1

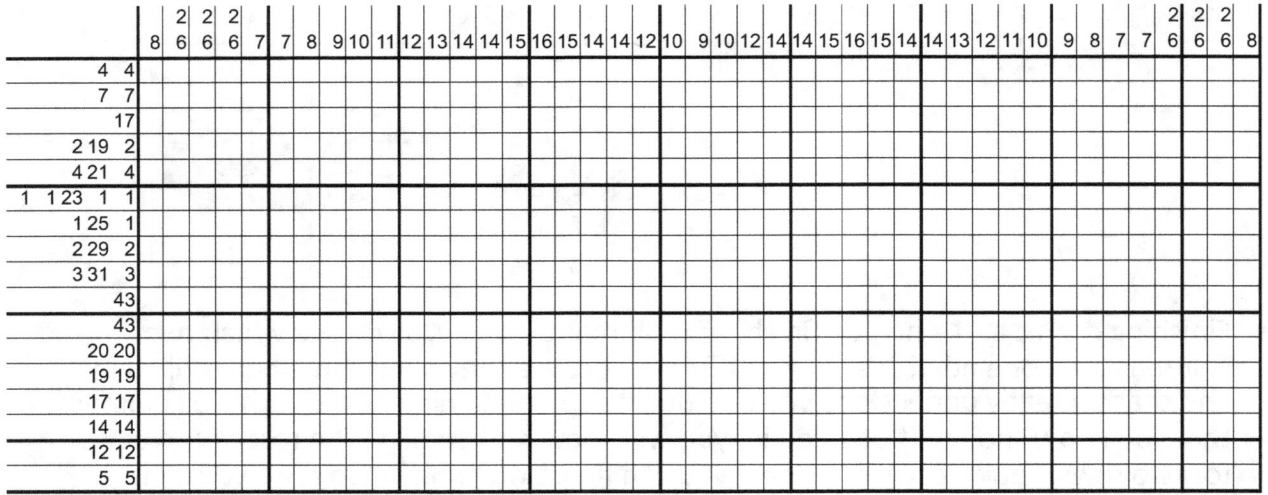

#2

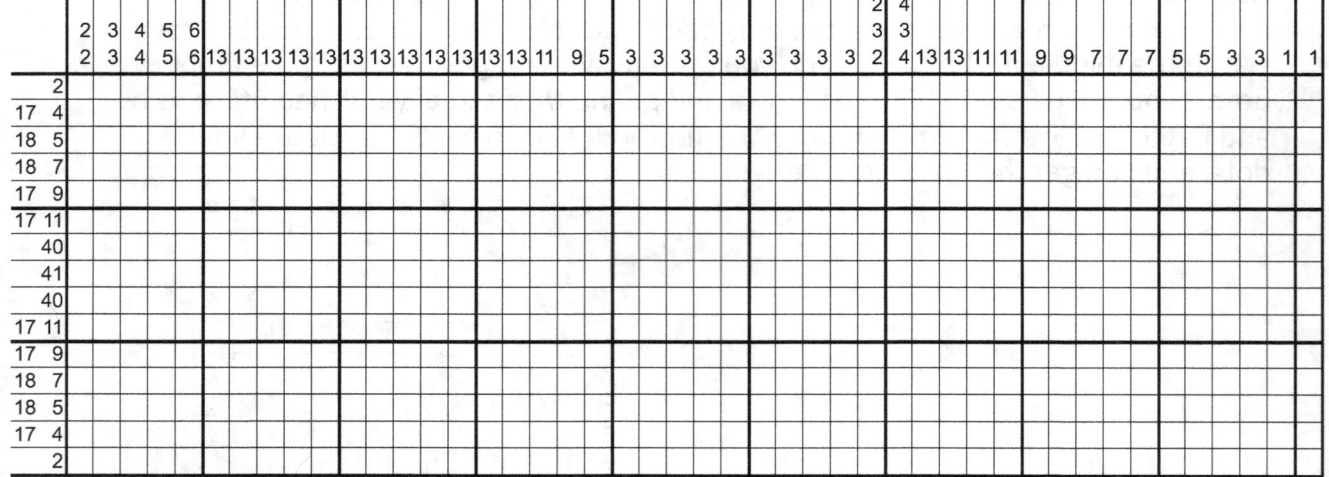

#3

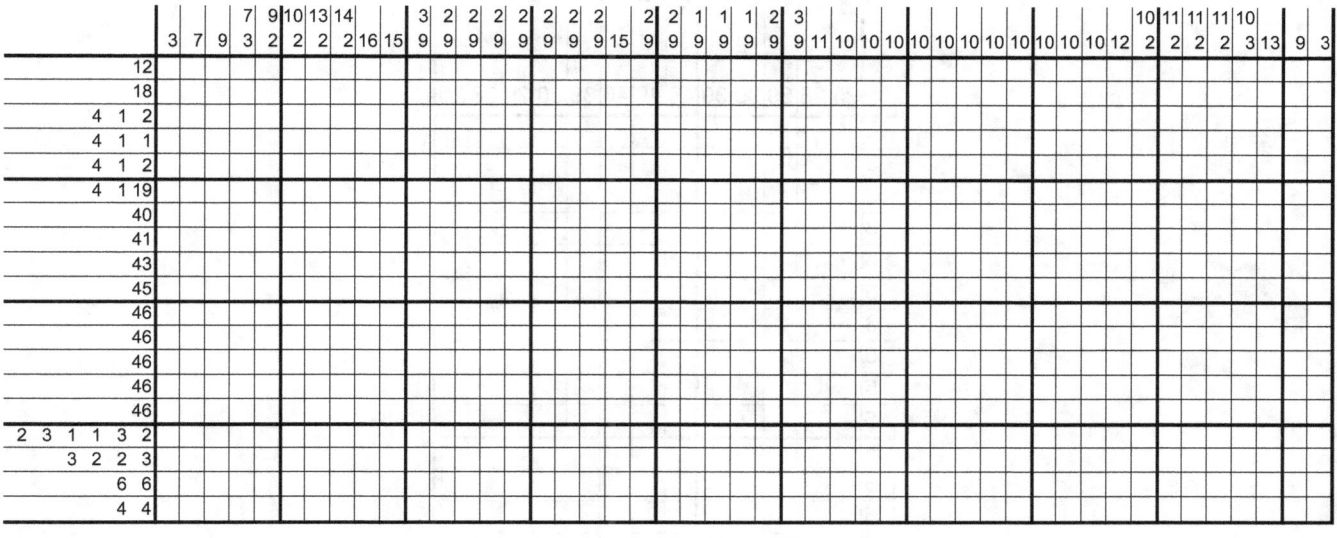

#4

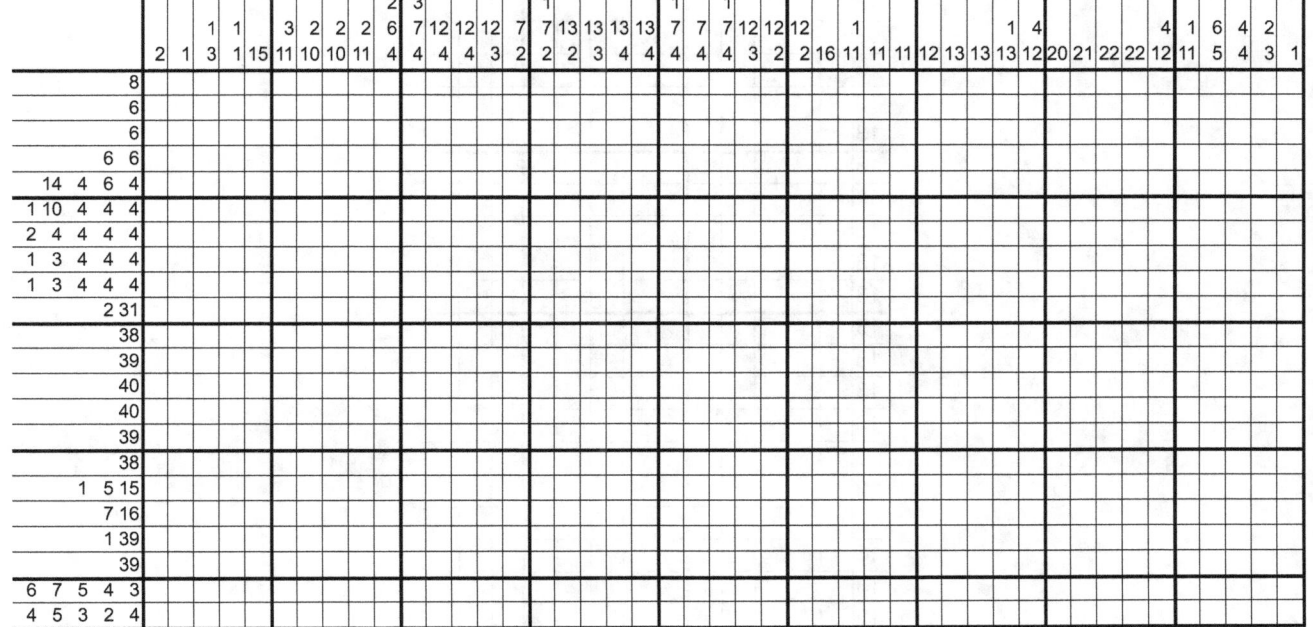

#5

Column clues (left to right)
| 25 | 28 | 29 | 30 | 1,1,39 | 46 | 46 | 46 | 1,1,39 | 30 | 29 | 28 | 25 |

Row clues (top to bottom)
5, 3, 3, 3, 3, 5, 3, 5, 5, 5, 5, 5, 5, 5, 5, 5, 7, 9, 11, 11, 13, 13, 13, 13, 13, 13, 13, 13, 13, 13, 13, 13, 13, 13, 13, 13, 13, 13, 13, 13, 13, 13, 13, 13, 13, 11

#6

Column clues (left to right): 12, 14, 14, 12, 12, 10, 10, 8, 8, 6, 6, 5, 5, 11/16, 43, 43, 44, 44, 5/16

Row clues (top to bottom):
- 2
- 5
- 7
- 9, 2
- 19
- 19
- 19
- 19
- 19
- 11, 5
- 9, 5
- 7, 5
- 5, 5
- 2, 5
- 5
- 4
- 4
- 4
- 4
- 4
- 4
- 4
- 4
- 4
- 4
- 4
- 4
- 4
- 4
- 4
- 4
- 6
- 6
- 6
- 6
- 6
- 6
- 6
- 6
- 6
- 6
- 6
- 6
- 6
- 6
- 6
- 6

#7

Nonogram puzzle

Column clues (left to right):
Col	Clues
1	4
2	10
3	12
4	13, 3, 3
5	14, 8
6	4, 8, 8
7	3, 8, 8
8	3, 9, 8
9	3, 9, 3, 8
10	3, 41
11	3, 41
12	3, 41
13	3, 41
14	3, 9
15	3, 9
16	3, 8
17	4, 8
18	14
19	13
20	12
21	10
22	4

Row clues (top to bottom):
18
20
20
5 5
5 5
5 5
6 6
22
20
20
20
18
16
14
10
8
4
4
4
4
4
4
4
4
4
4
4
4
4
4
4
4
4
4
4
5
5
5
4
4
10
10
10
9
9
10
10
10
4
4

#8

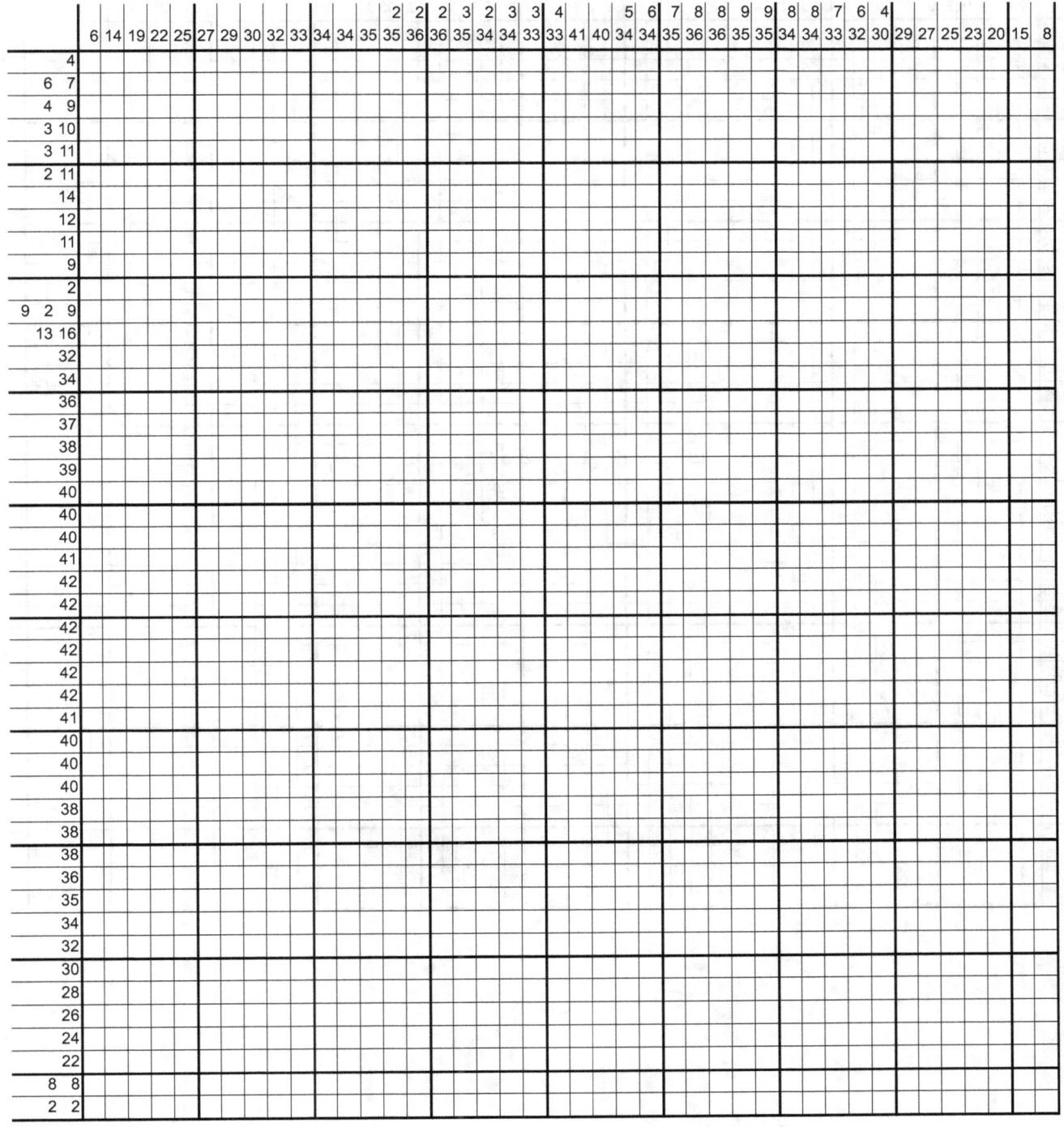

#9

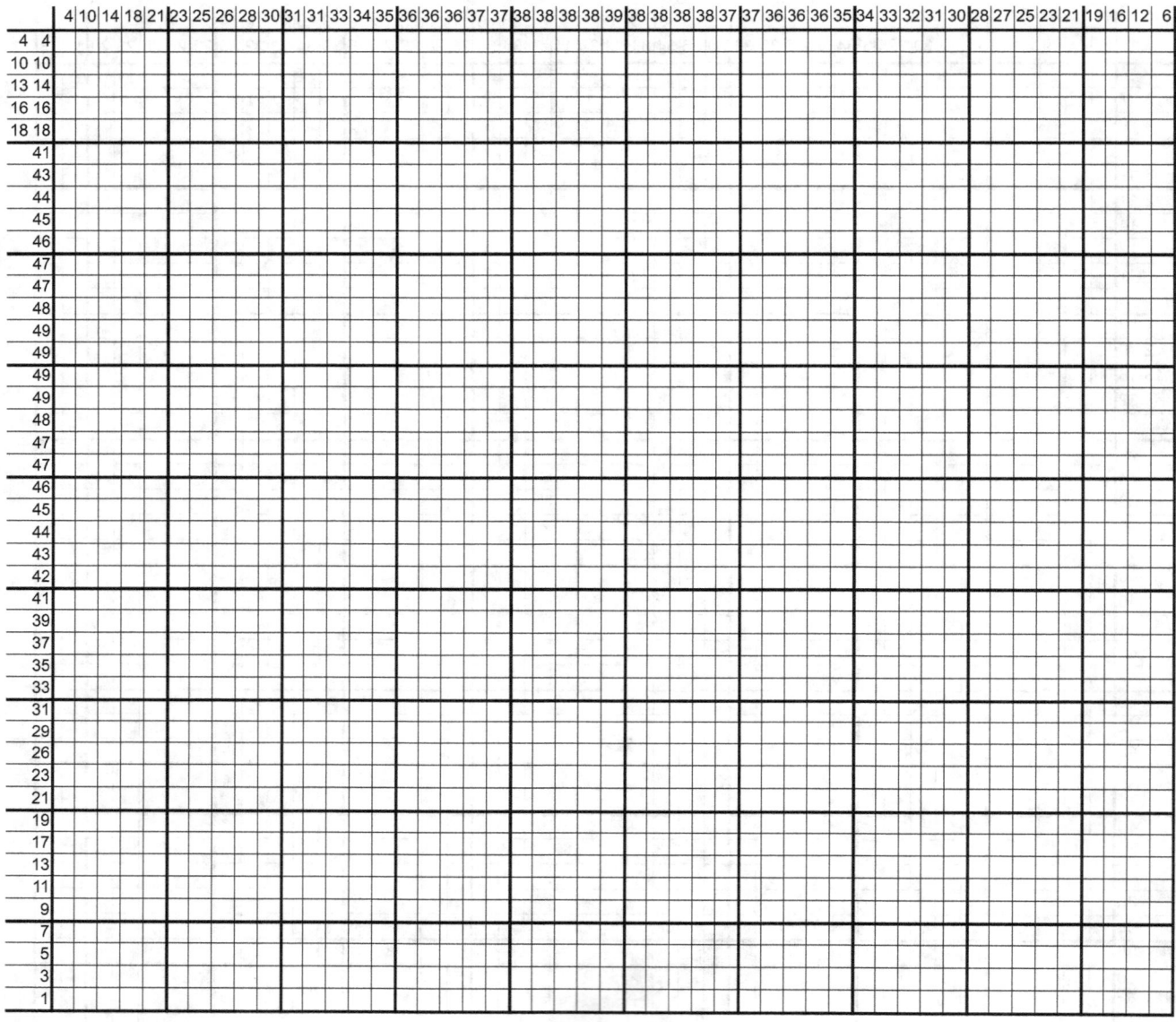

#10

	11	16	19	23	30	40	43	45	46	46	38	33	28	25	23	24	25	29	33	38	46	46	45	43	40	30	23	19	16	11
8 9																														
11 12																														
26																														
28																														
28																														
30																														
30																														
30																														
30																														
30																														
30																														
30																														
30																														
30																														
30																														
28																														
28																														
28																														
26																														
26																														
24																														
24																														
24																														
22																														
10 10																														
9 9																														
9 9																														
9 9																														
8 8																														
7 7																														
7 7																														
7 7																														
6 6																														
6 6																														
6 6																														
6 6																														
6 6																														
5 5																														
5 5																														
4 4																														
4 4																														
4 4																														
3 3																														
3 3																														
2 2																														

#11

Column clues (left to right):

								4	5	4								
						1		17	17	17		1						
1	3	29	31	33	35	37	41	13	13	13	41	37	35	33	31	29	3	1

Row clues (top to bottom):

1
3
5
7
5
1 1
1 1
1 1
5
5
5
7
7
9
9
11
11
13
13
15
15
17
17
19
15
6 6
6 6
6 6
6 6
15
15
15
15
15
15
15
15
15
15
15
15
15
6 6
5 5
5 5
5 5
5 5
5 5

#12

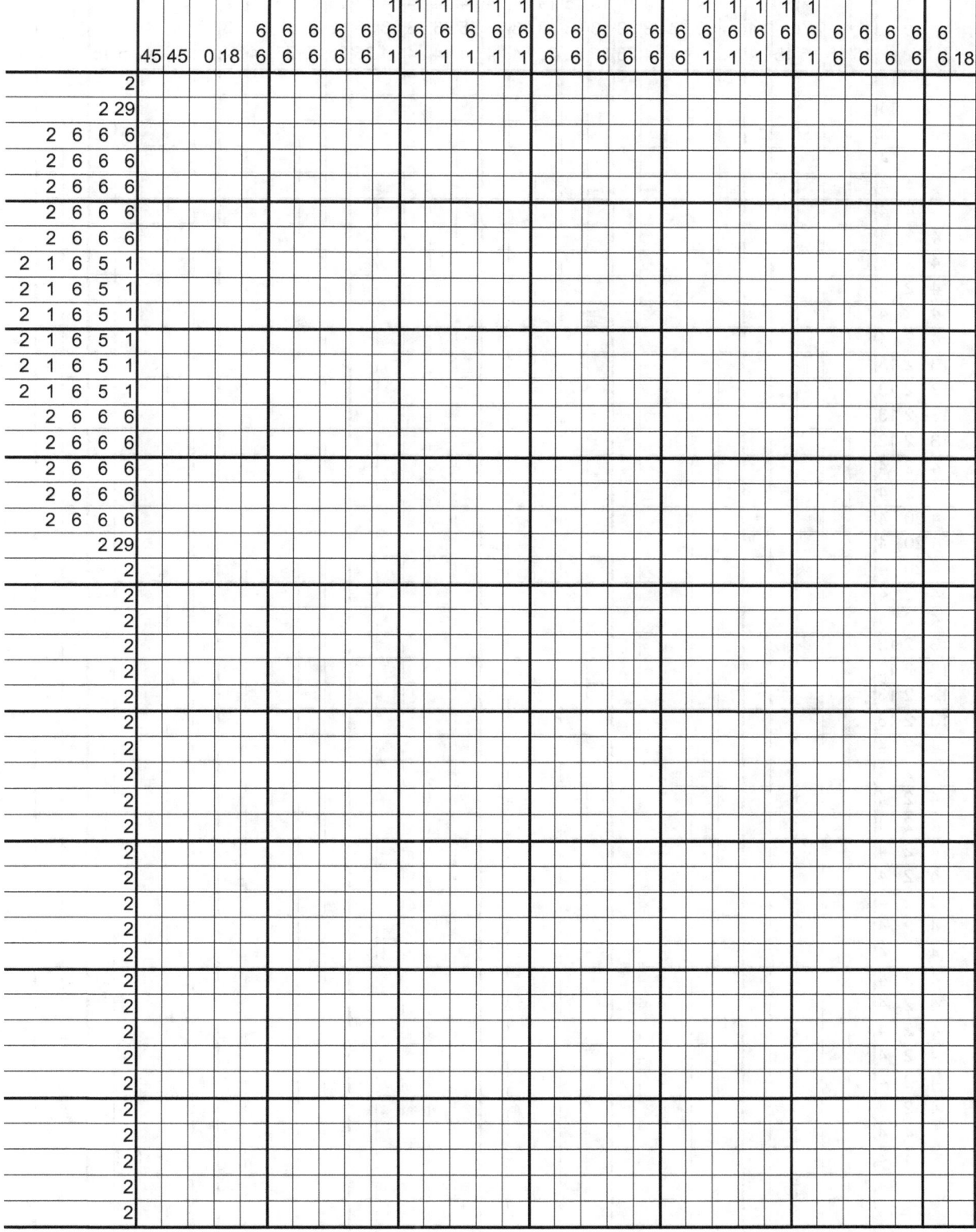

#13

Column clues (left to right)
15, 15, 15, 15, 4, 4/4, 4/3/4, 4/3/4, 4/15/3/25, 15/3/25, 15/3/25, 15/3/25, 4/3/4, 4/4, 4/35/4, 4/35/4, 4/3/4, 4/15/3/25, 15/3/25, 15/3/25, 15/3/25, 4/3/4, 4/3/4, 4/3/4, 4/4, 4/4, 15, 15, 15, 15

Row clues (top to bottom)
- 14
- 14
- 14
- 14
- 4 4
- 4 4
- 4 2 4
- 4 2 4
- 4 2 4
- 4 2 4
- 4 2 4
- 13 2 13
- 13 2 13
- 13 2 13
- 13 2 13
- 4 2 4
- 4 2 4
- 4 20 4
- 4 20 4
- 4 20 4
- 4 2 4
- 4 2 4
- 13 2 13
- 13 2 13
- 13 2 13
- 13 2 13
- 4 2 4
- 4 2 4
- 4 2 4
- 4 2 4
- 4 2 4
- 4 2 4
- 4 2 4
- 4 2 4
- 4 2 4
- 4 2 4
- 4 2 4
- 4 2 4
- 4 2 4
- 4 4
- 4 4
- 14
- 14
- 14
- 14

#14

Column clues:

											4	3	3	3	3	3	3	3	4											
						8	5	5	5	4	9	8	7	7	7	7	7	8	9	4	5	5	5	8						
22	24	28	31	33	35	25	25	25	25	25	14	13	6	6	6	6	6	13	14	25	25	25	25	25	35	33	31	28	24	22

Row clues:

	9
	13
	15
6	6
6	6
5	5
5	5
4	4
4	4
4	4
4	4
4	4
4	4
4	4
	29
	31
	31
	31
	31
	31
	31
13	13
12	12
11	11
11	11
12	12
13	13
13	13
13	13
13	13
13	13
13	13
13	13
	31
	31
	31
	31
	29
	27

#15

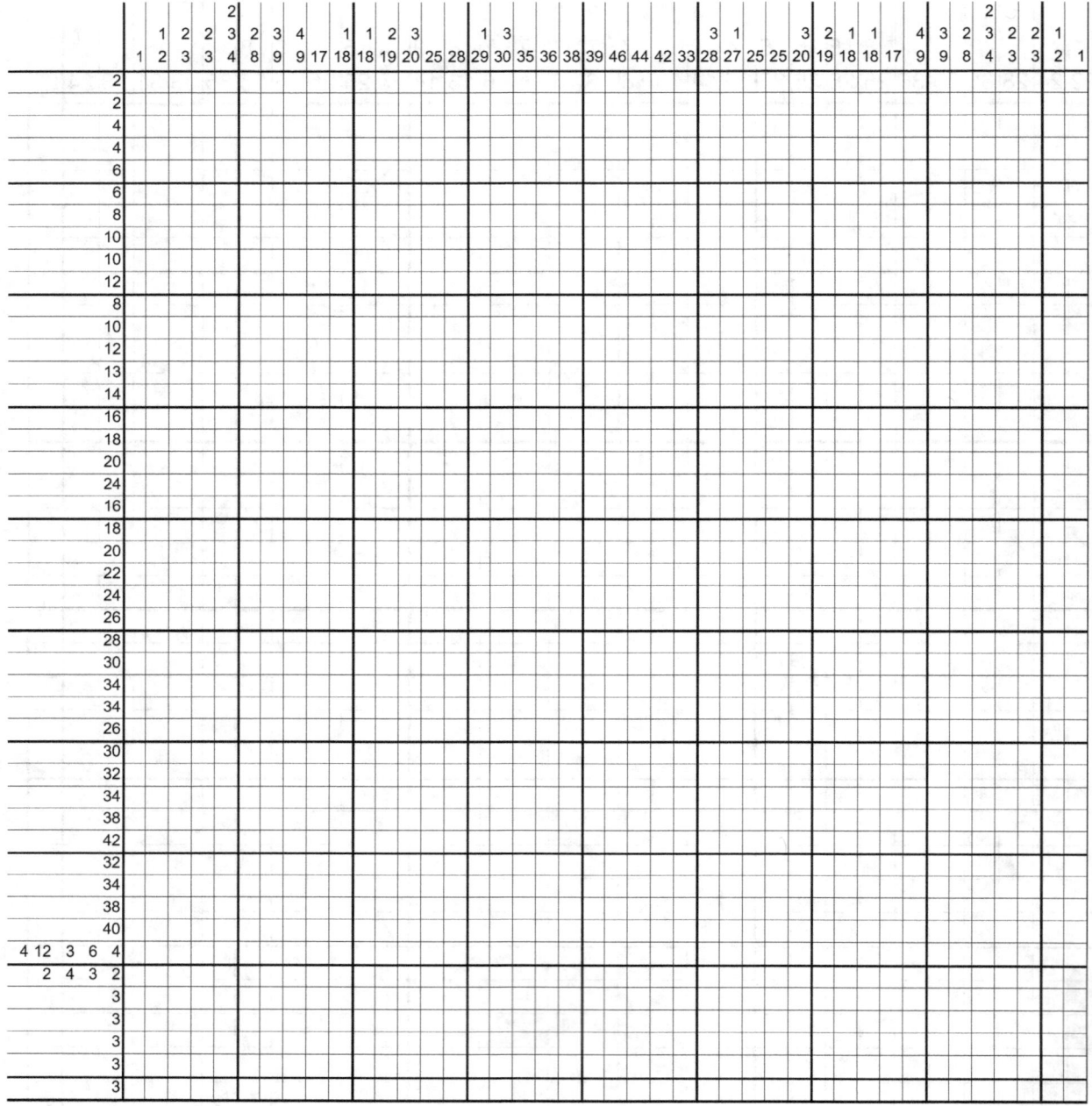

#16

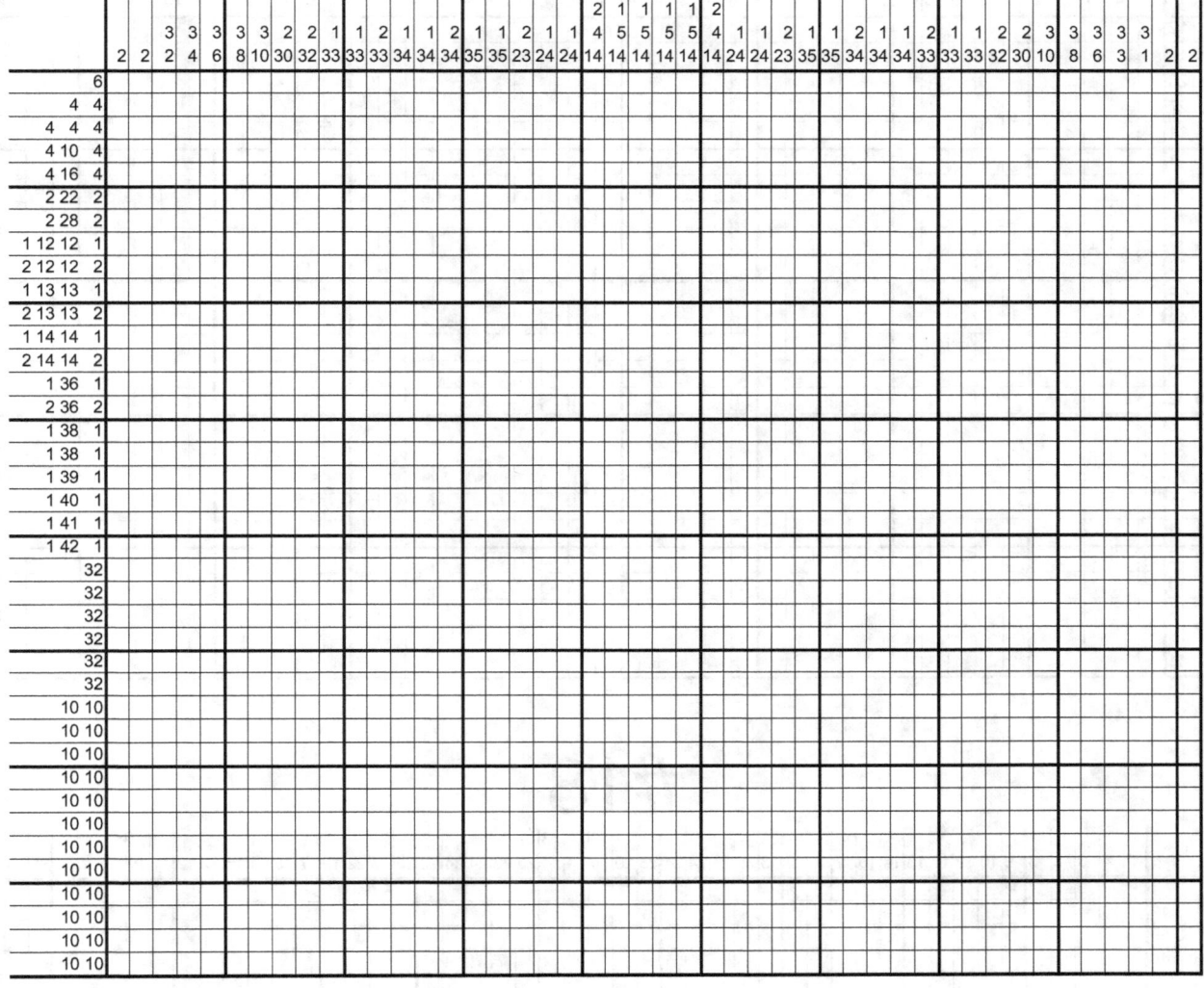

#17

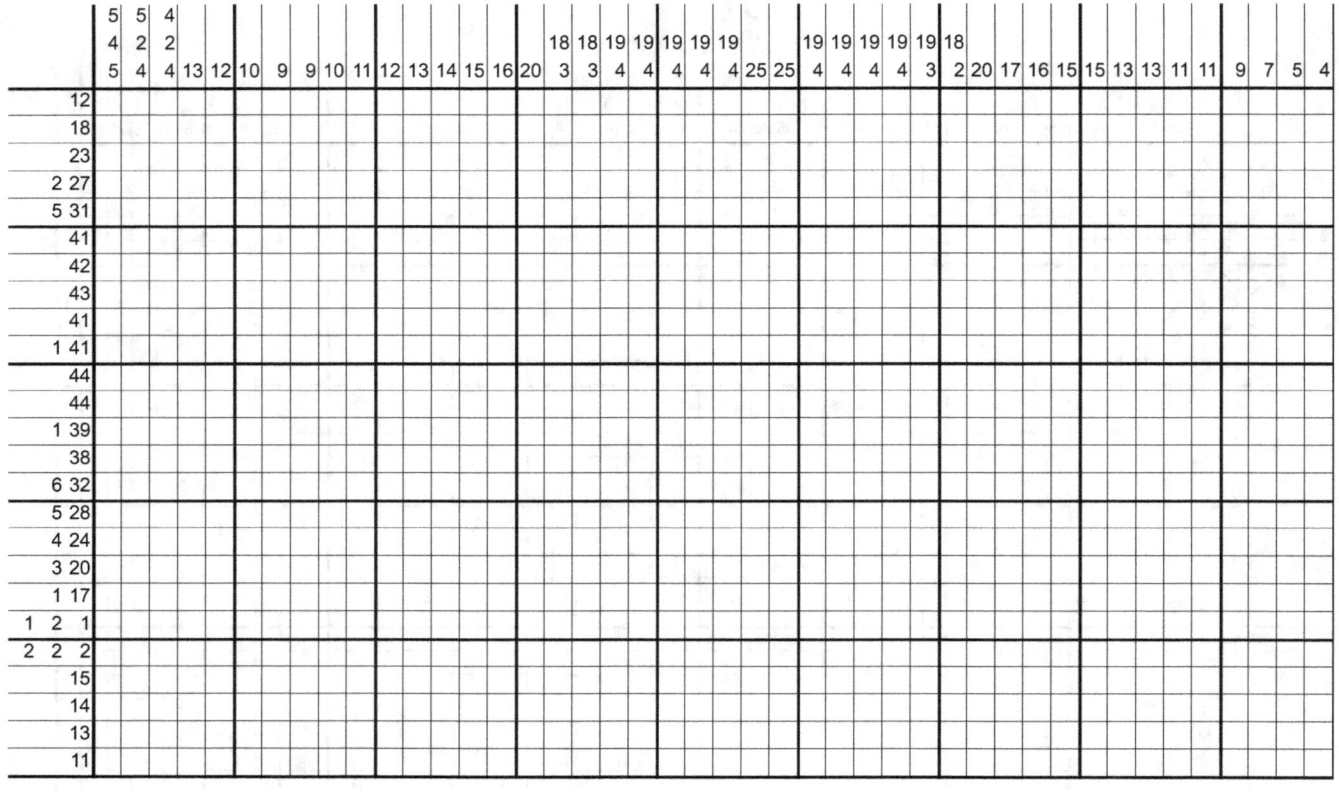

#18

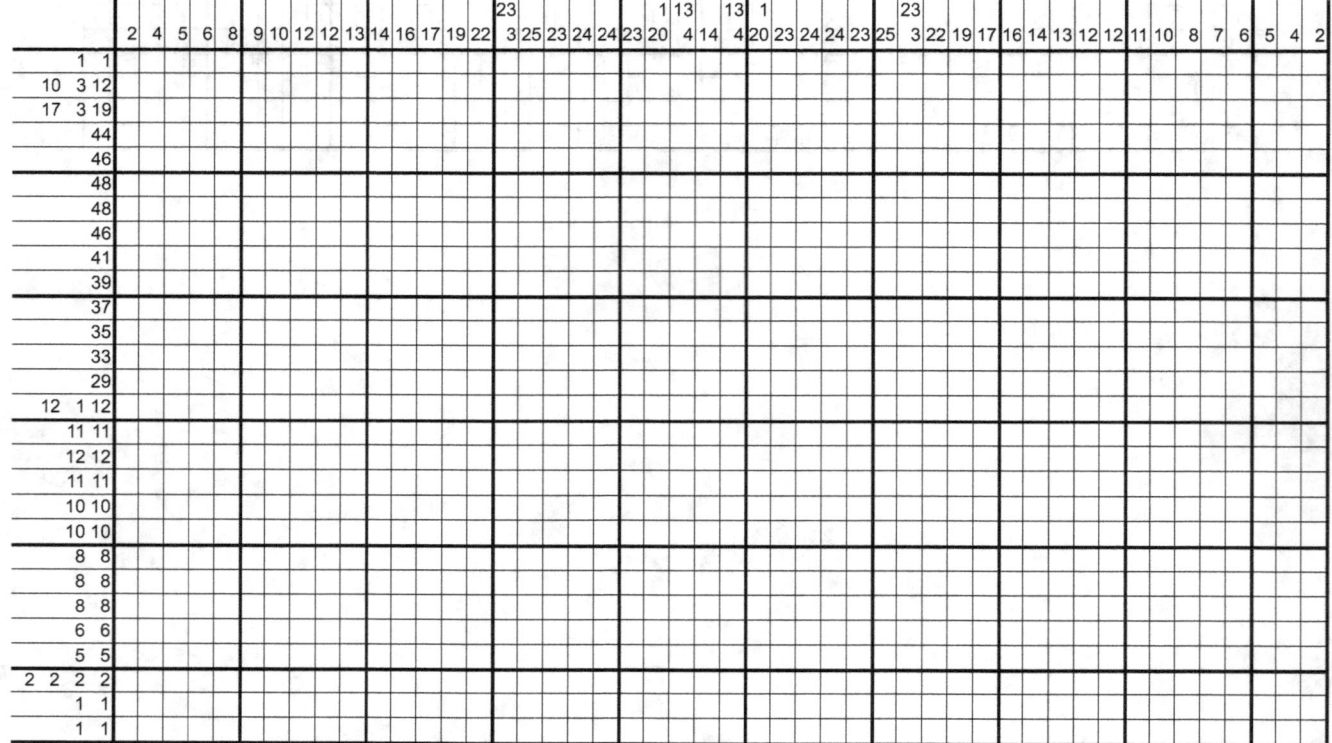

#19

Column clues (left to right)

									1											
					2	1	1	4	4	1	4		4	1	4					
26	28	30	30	30	30	30	30	30	31	37	38	46	46	38	37	31	30	30	30	30	30	30	28	26

Row clues (top to bottom)

- 10
- 1 4
- 4
- 4
- 2
- 2
- 2
- 2
- 4
- 6
- 8
- 8
- 8
- 8
- 6
- 8
- 21
- 23
- 25
- 25
- 25
- 25
- 25
- 25
- 25
- 25
- 25
- 25
- 25
- 25
- 25
- 25
- 25
- 25
- 25
- 25
- 25
- 25
- 25
- 25
- 25
- 25
- 23
- 21

#20

Column clues (left to right)

Col	Clues
1	2
2	4
3	5
4	6
5	7
6	3,4
7	3,4,1
8	3,4,1
9	3,4,2
10	3,3,4,2
11	3,5,4,2
12	3,2,11,2
13	3,2,9,2
14	3,2,10,2
15	3,7,25
16	3,6,25
17	3,6,25
18	3,7,6,2
19	3,13,2
20	15,2
21	2,3,2,7,2
22	7,2,6,2
23	7,8,2
24	7,7,1
25	2,3,5,1
26	3,4
27	7
28	6
29	5
30	3
31	2

Row clues (top to bottom)

1. 3
2. 5
3. 5
4. 3
5. 31
6. 31
7. 29
8. 4,1,3
9. 4,20
10. 3,19
11. 3,3,6,5
12. 3,2,12
13. 3,15
14. 4,13
15. 8,7
16. 6,5
17. 5,5
18. 11
19. 9
20. 7
21. 7
22. 5
23. 3
24. 3
25. 3
26. 3
27. 3
28. 3
29. 3
30. 3
31. 3
32. 3
33. 3
34. 3
35. 3
36. 3
37. 3
38. 15
39. 19

#21

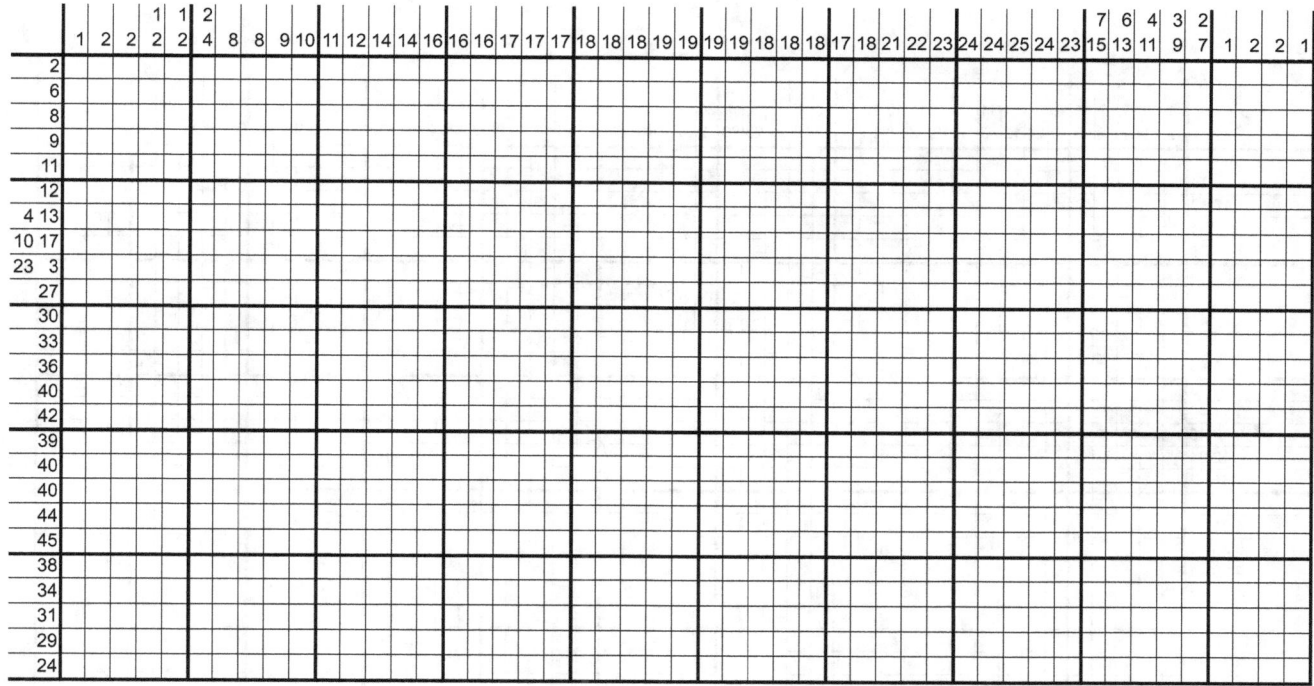

#22

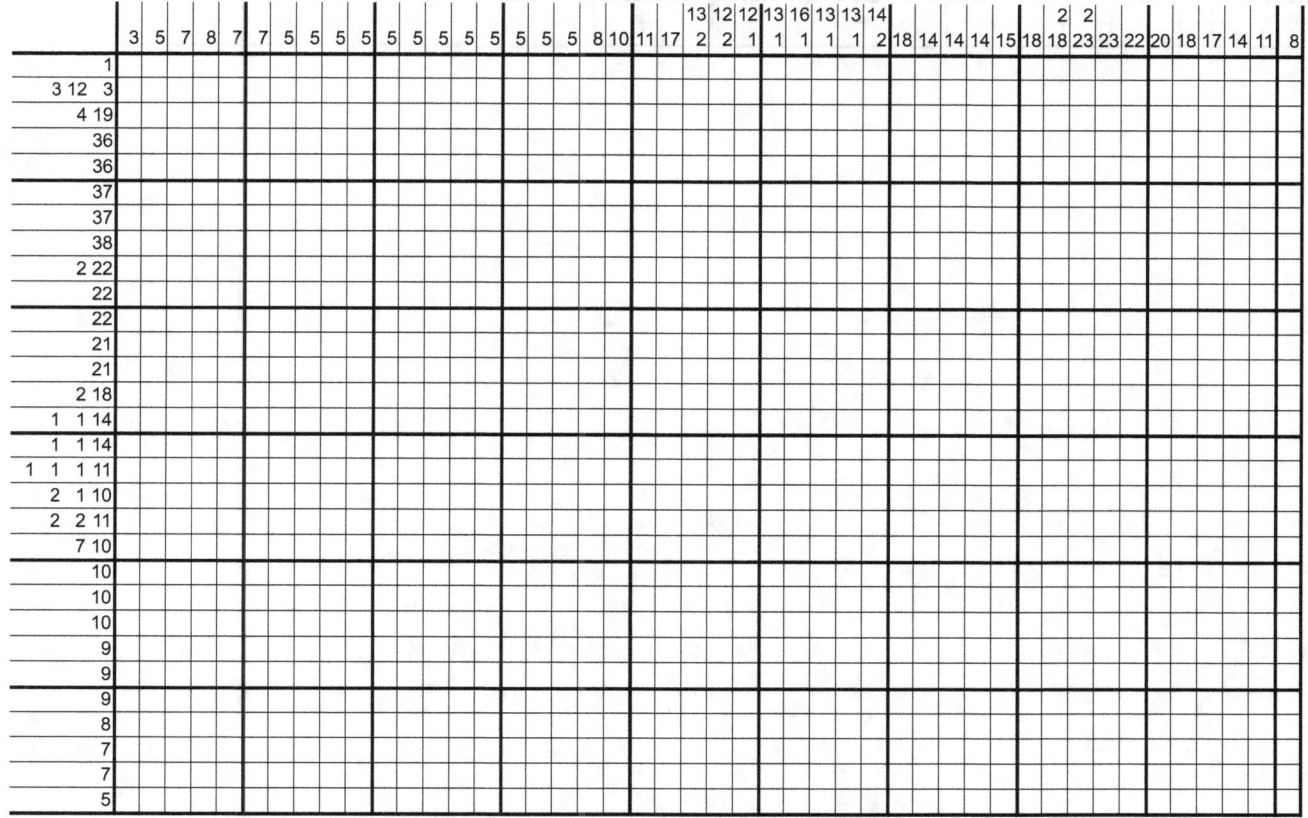

#23

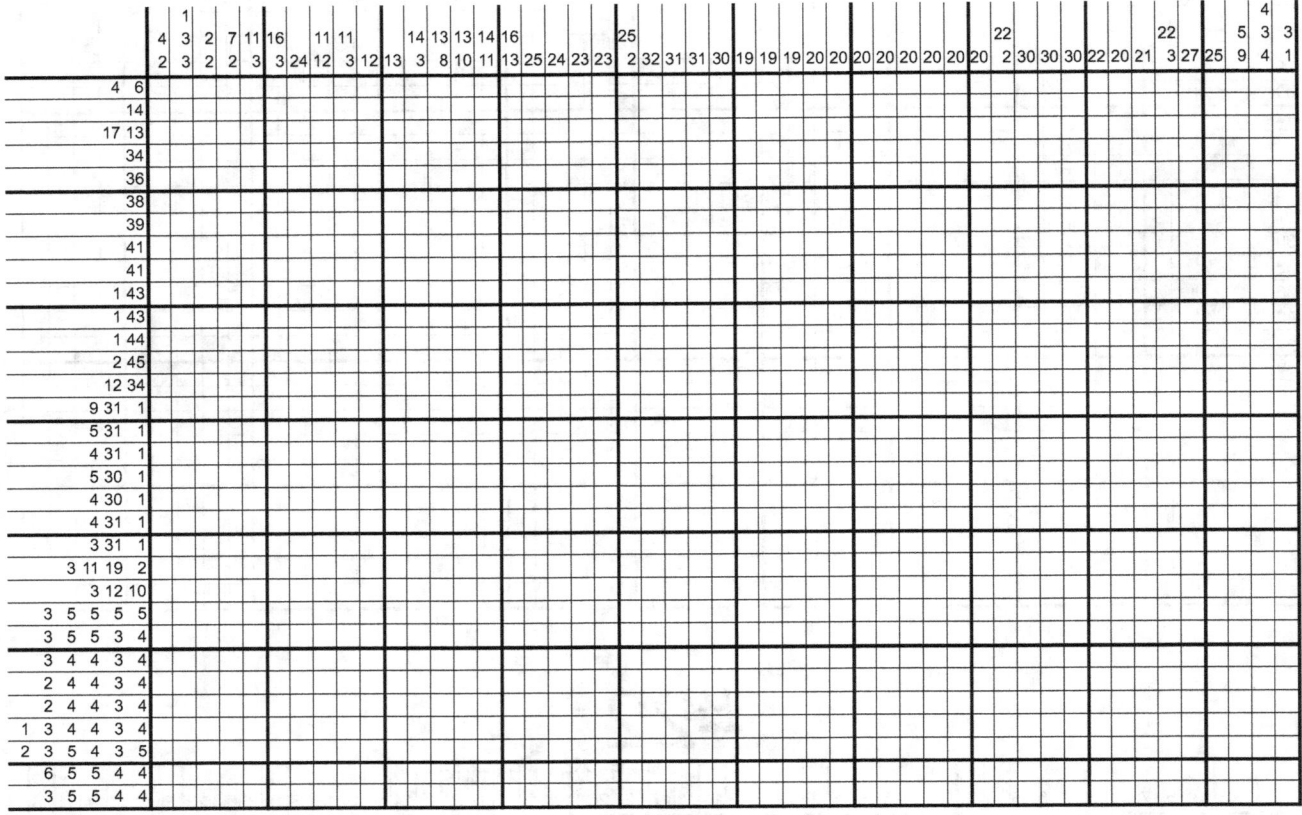

#24

Column clues (left to right):

1. 1
2. 3 2
3. 1 5 3
4. 2 5 6
5. 2 5 8
6. 6 5 8
7. 40
8. 41
9. 16 24
10. 15 22
11. 15 22
12. 17 1 21
13. 18 3 21
14. 17 1 21
15. 15 22
16. 15 22
17. 16 24
18. 41
19. 40
20. 6 5 8
21. 2 5 8
22. 2 5 6
23. 1 5 3
24. 3 2
25. 1

Row clues (top to bottom):

1. 1
2. 3
3. 3
4. 3
5. 7
6. 9
7. 11
8. 13
9. 13
10. 15
11. 15
12. 15
13. 15
14. 21
15. 19
16. 13
17. 13
18. 13
19. 5 5
20. 3 3
21. 3 3
22. 6 1 6
23. 7 3 7
24. 9 1 9
25. 8 8
26. 9 9
27. 13
28. 13
29. 13
30. 13
31. 13
32. 13
33. 13
34. 13
35. 13
36. 13
37. 13
38. 13
39. 13
40. 17
41. 19
42. 19
43. 19
44. 21
45. 23
46. 23
47. 17

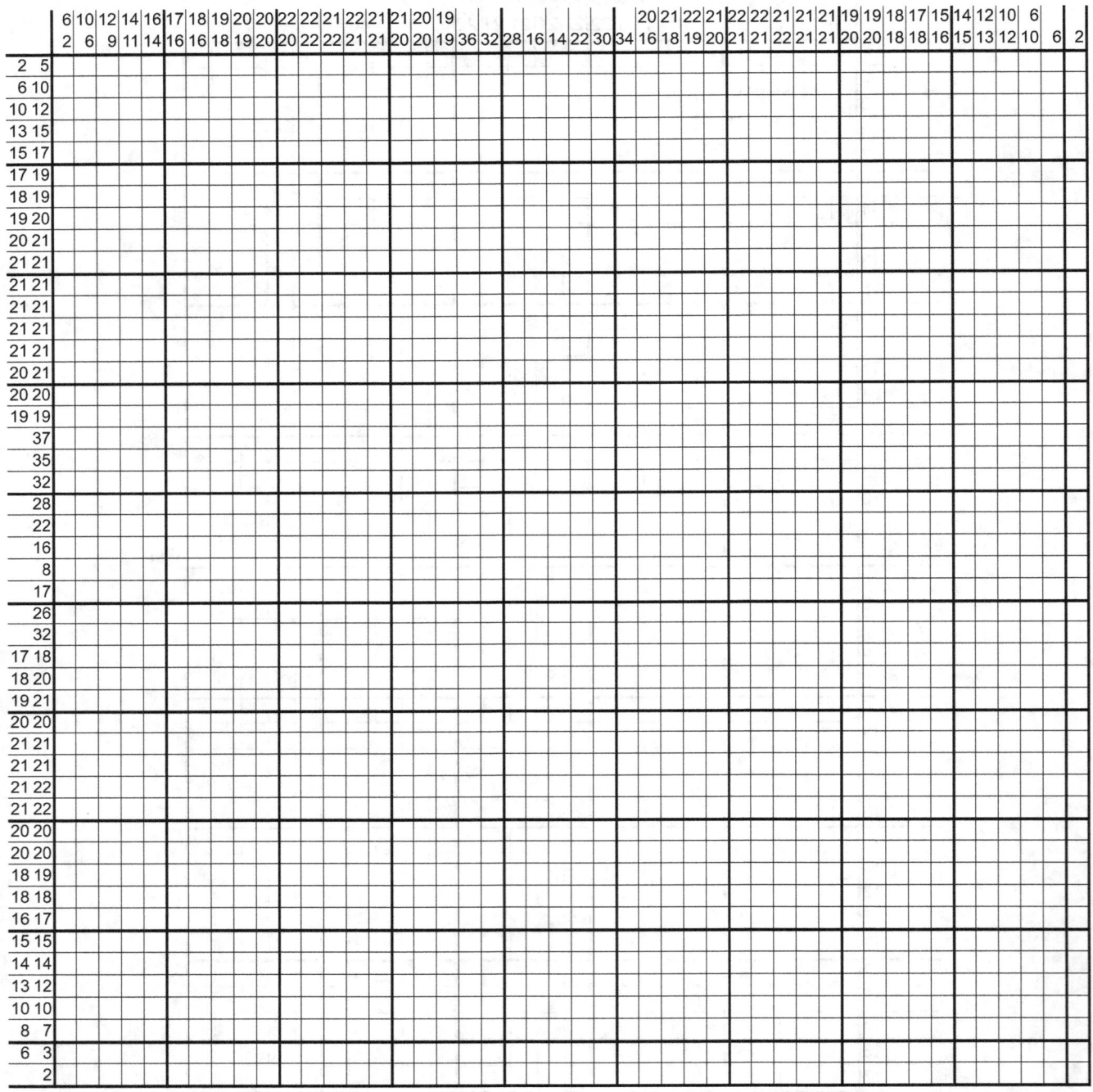

#26

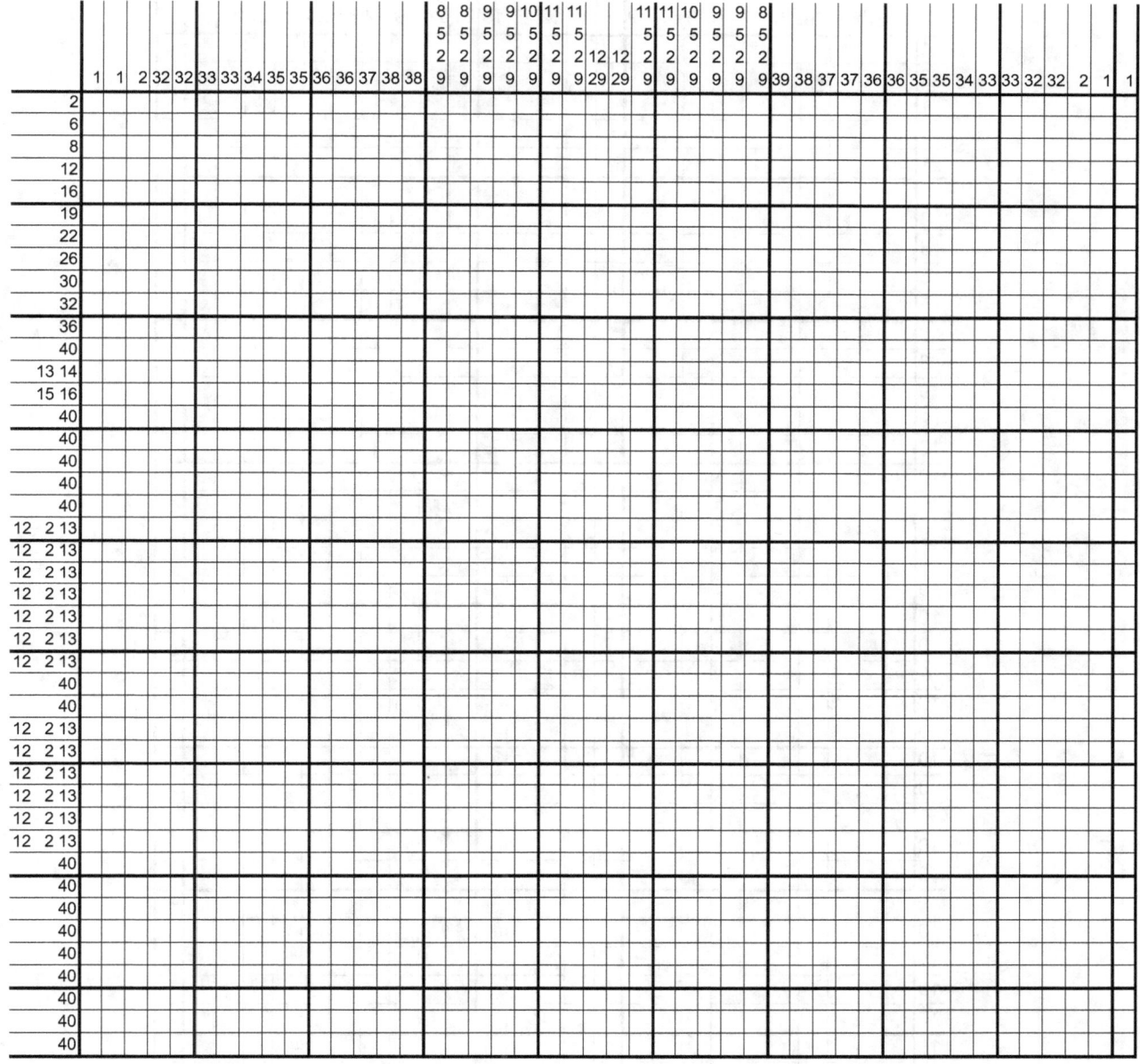

#27

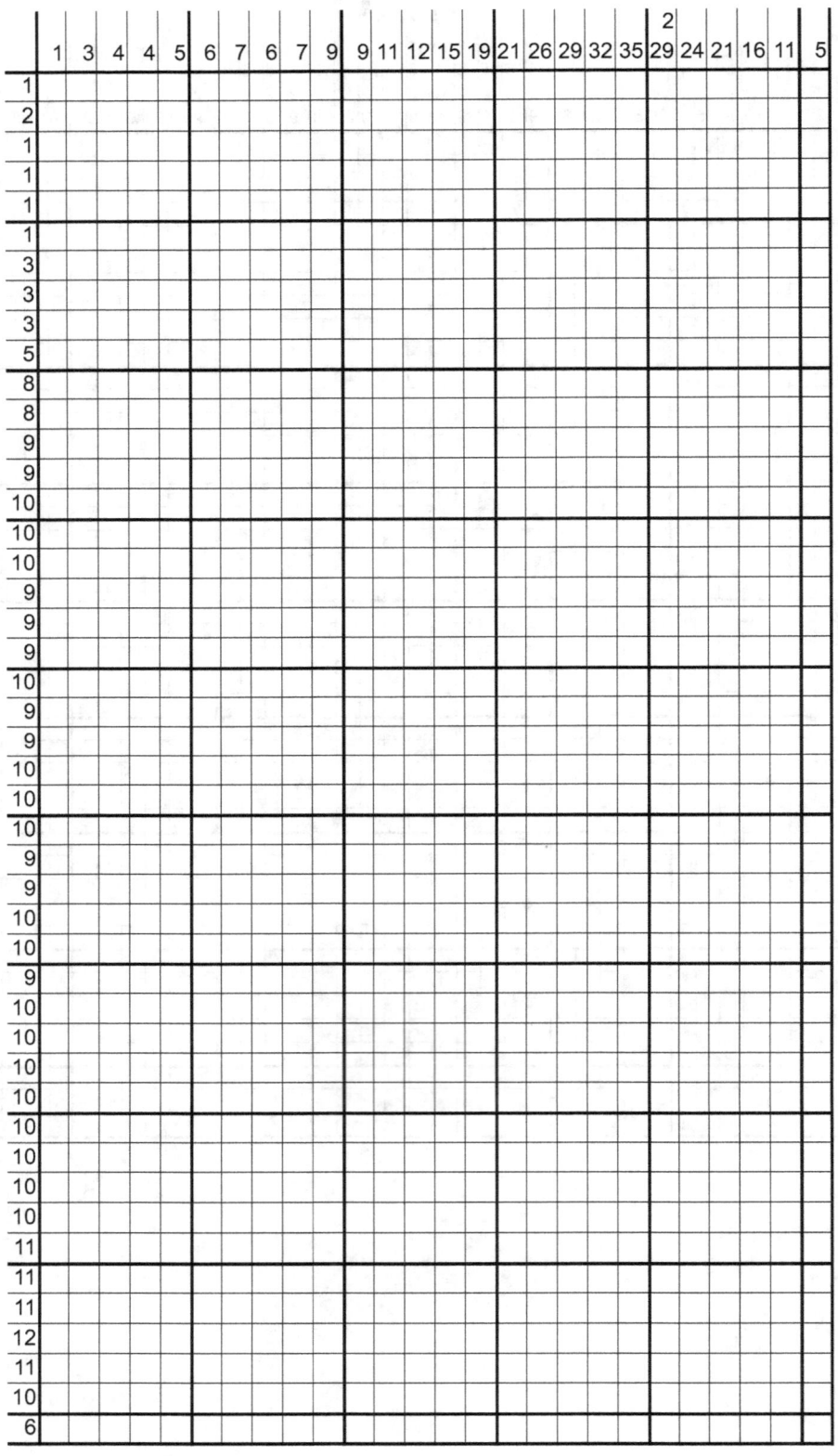

#28

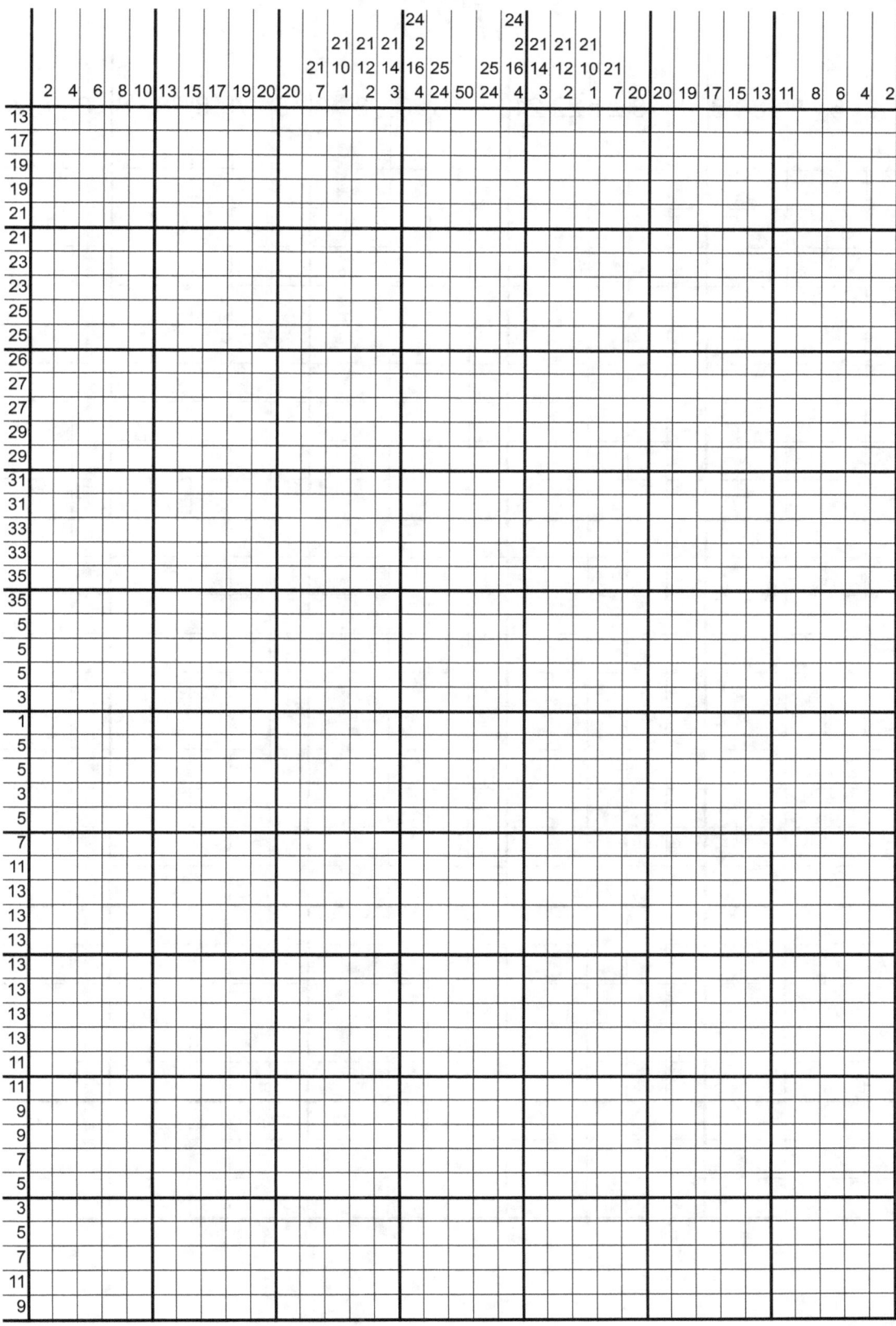

#29

	4	6	8	8	8	8	8	21	22	21	8	8	7	8	8/3	7/5	8/7	8/8	7/8	8/8	8/8	8/8	24	23	22
3																									
6																									
9																									
12																									
15																									
17																									
17																									
18																									
18																									
11 3																									
8 3																									
5 3																									
3 3																									
3 3																									
3 3																									
3 3																									
3 3																									
3 8																									
3 9																									
3 10																									
8 11																									
9 11																									
10 10																									
10 8																									
10 6																									
10																									
8																									
6																									

#30

Nonogram puzzle grid.

Column clues (left to right):
17 | 23 | 24 | 33 | 34 | 16,17 | 16,17 | 16,17 | 15,18 | 15,18 | 15,19 | 15,19 | 15,20 | 15,6,13 | 16,7,14 | 28,14 | 28,15 | 28,15 | 28,15 | 28,15 | 28,14 | 16,7,14 | 15,6,13 | 15,20 | 15,19 | 15,19 | 15,18 | 15,18 | 16,17 | 16,17 | 16,17 | 34 | 33 | 24 | 23 | 17

Row clues (top to bottom):
- 2, 2
- 8, 8
- 36
- 36
- 36
- 36
- 36
- 36
- 36
- 36
- 36
- 36
- 36
- 36
- 36
- 36
- 4, 8, 4
- 4, 6, 4
- 6, 6, 6
- 8, 6, 8
- 10, 6, 10
- 34
- 32
- 32
- 30
- 30
- 30
- 10, 8, 10
- 10, 6, 10
- 12, 12
- 30
- 30
- 30
- 28
- 26
- 22
- 20
- 18
- 16
- 14
- 12
- 10
- 8
- 6
- 4

#31

Column clues (left to right)

Col	Clues
1	14, 1
2	17, 1
3	17, 1
4	18, 2
5	17, 2
6	6, 2
7	6, 3
8	5, 3
9	44
10	46
11	47
12	47
13	46
14	45
15	4, 3
16	5, 3
17	10, 2
18	11, 2
19	12, 2
20	10, 1
21	8, 1
22	1

Row clues (top to bottom)

1. 2
2. 4
3. 5
4. 6
5. 6
6. 6
7. 6
8. 6
9. 6
10. 6
11. 6
12. 6
13. 3 6
14. 5 6 1
15. 5 6 3
16. 5 6 4
17. 5 6 5
18. 5 6 5
19. 5 6 5
20. 5 6 5
21. 5 6 5
22. 5 6 5
23. 5 6 6
24. 5 12
25. 5 11
26. 7 9
27. 16
28. 13
29. 13
30. 11
31. 9
32. 6
33. 6
34. 6
35. 6
36. 6
37. 6
38. 6
39. 6
40. 6
41. 6
42. 6
43. 6
44. 10
45. 16
46. 22

#32

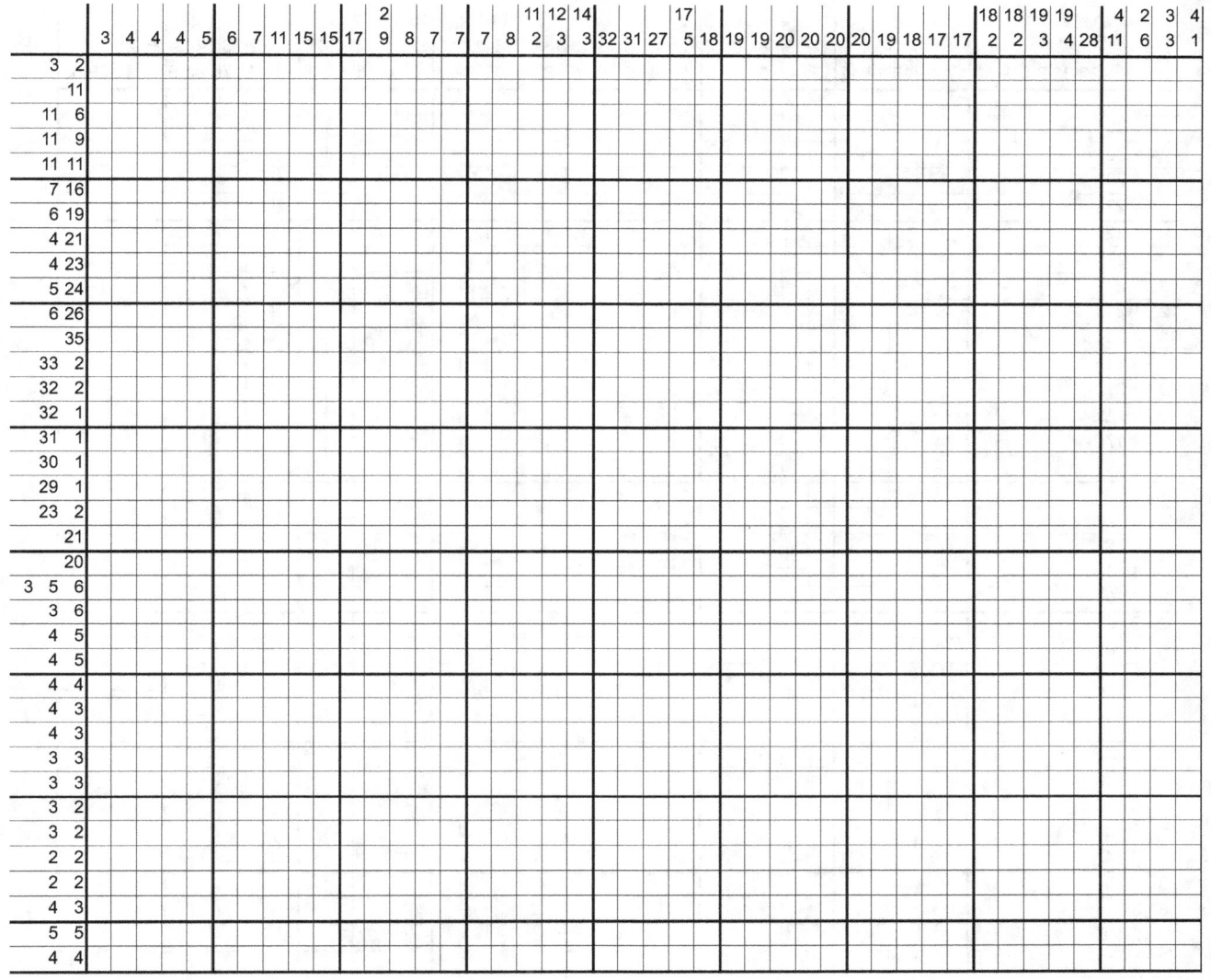

#33

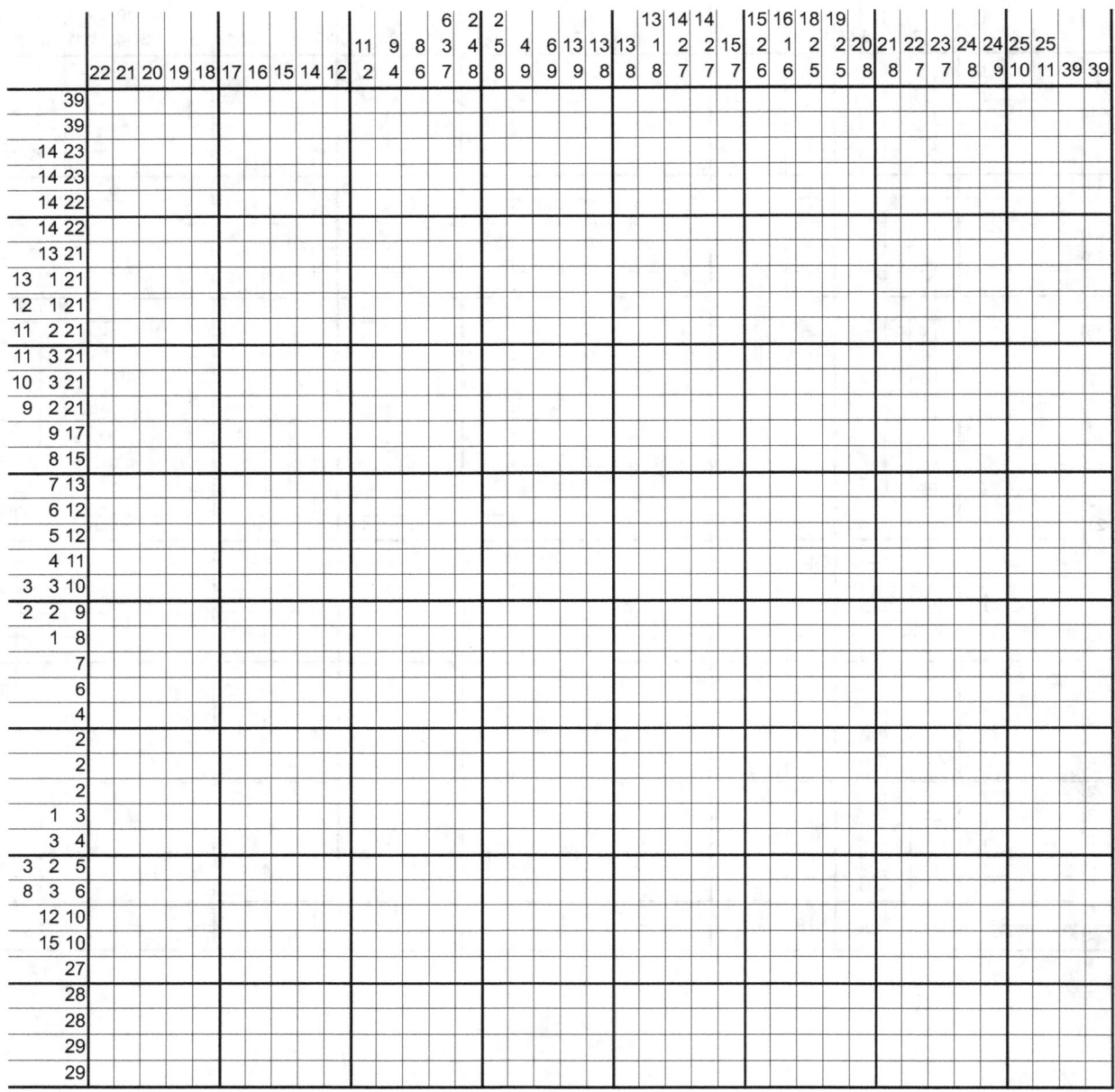

#34

Column clues (left to right, top to bottom):

7
5 10
23
23
24
8 10
1 1 1 1 9
30 1 1 1 8
31 1 1 1 8
31 1 1 1 9
1 1 1 2 9
1 9
1 9
2 8
1 5 6
4 2 5
3 4
6

Row clues (top to bottom):

3
5
3
3
5
3
3
5
3
3
3
3
3
3
3
3
3
3
3
3
3
3
3
3
3 3
4 3
4 3 3
5 3 1 1
10 1 1
5 5 1
4 2 1
4 1
4 3 1
4 1
4 3 1
3 1
4 2
4 3 1
6 2
7 4 1
14 1
15 1
18
18
17
16
14
10

#35

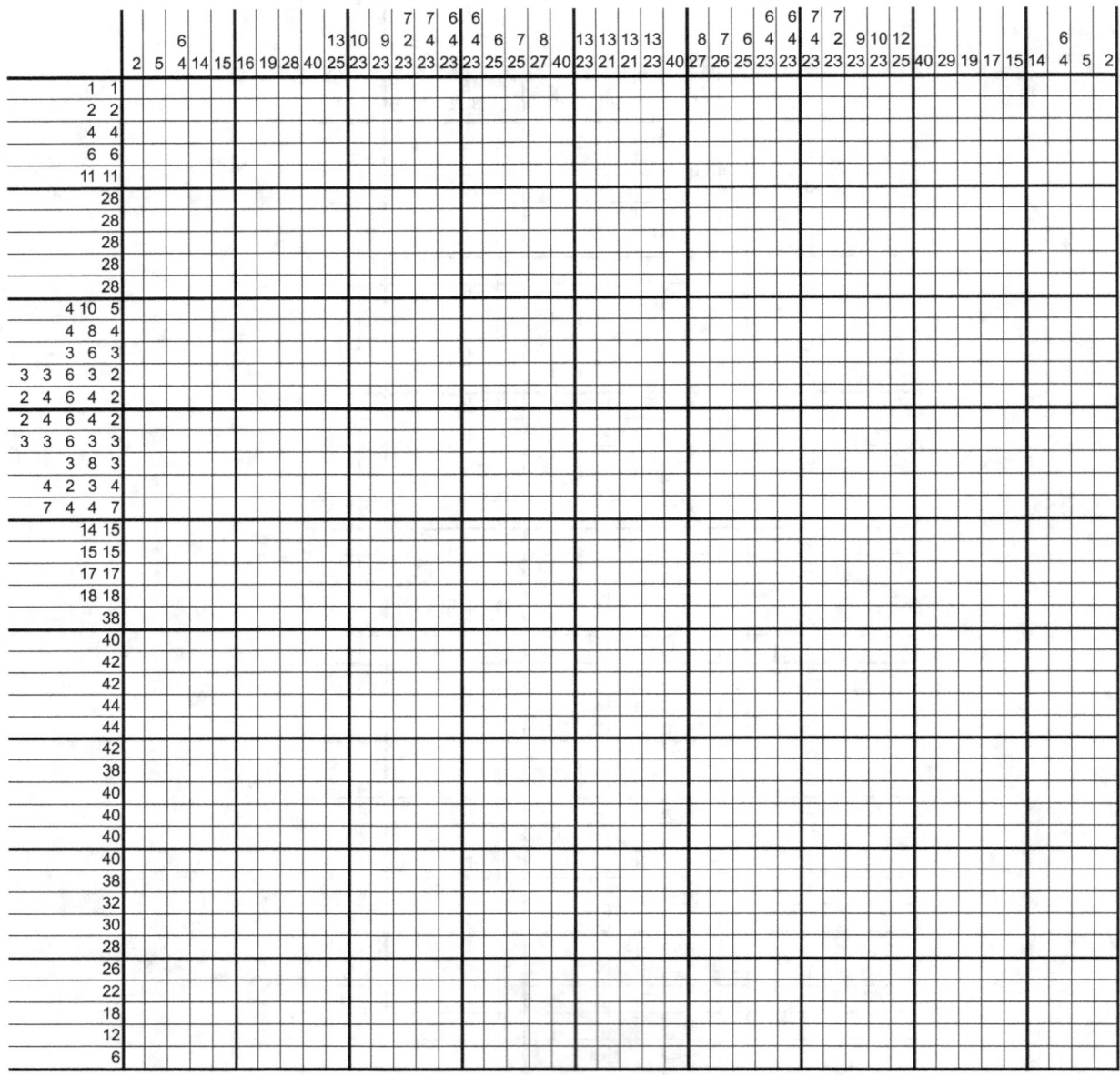

#36

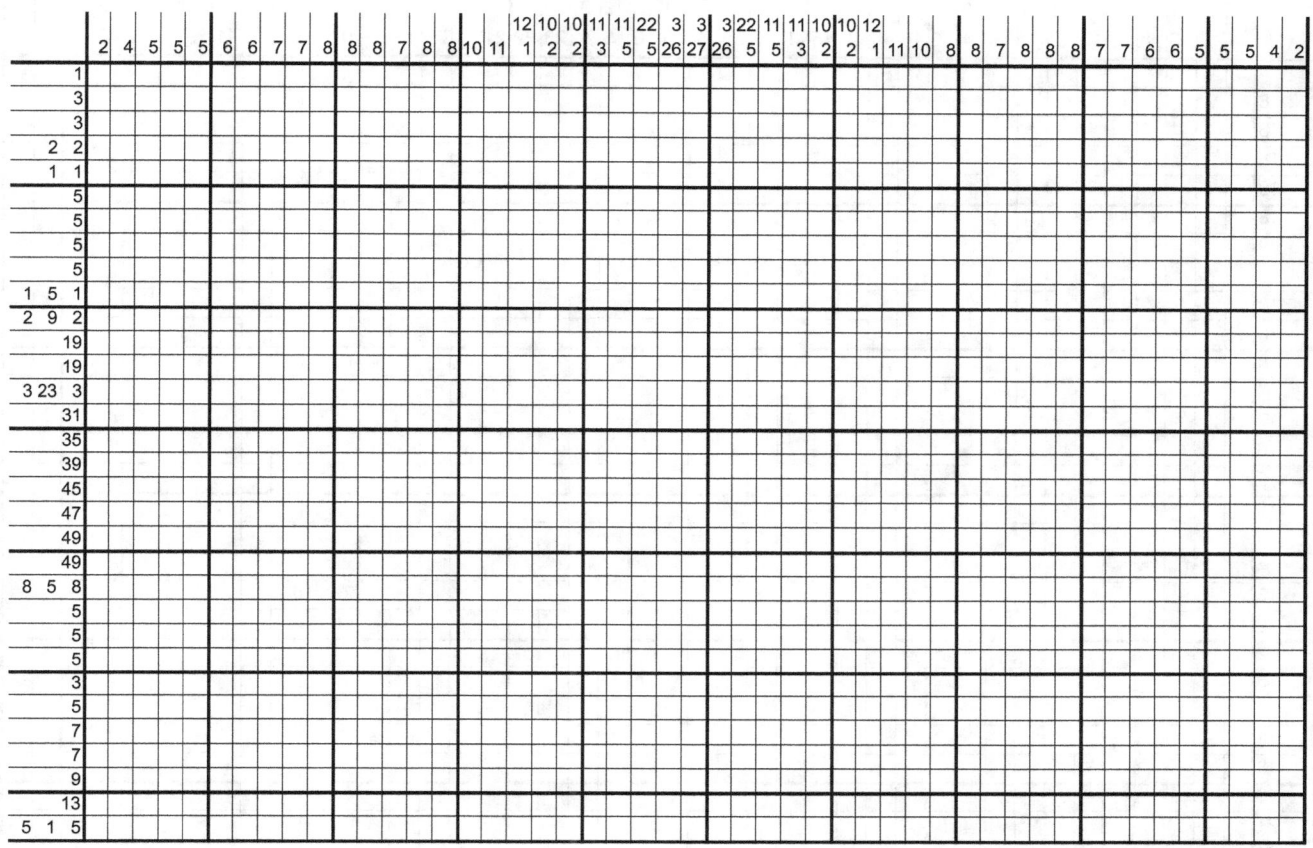

#37

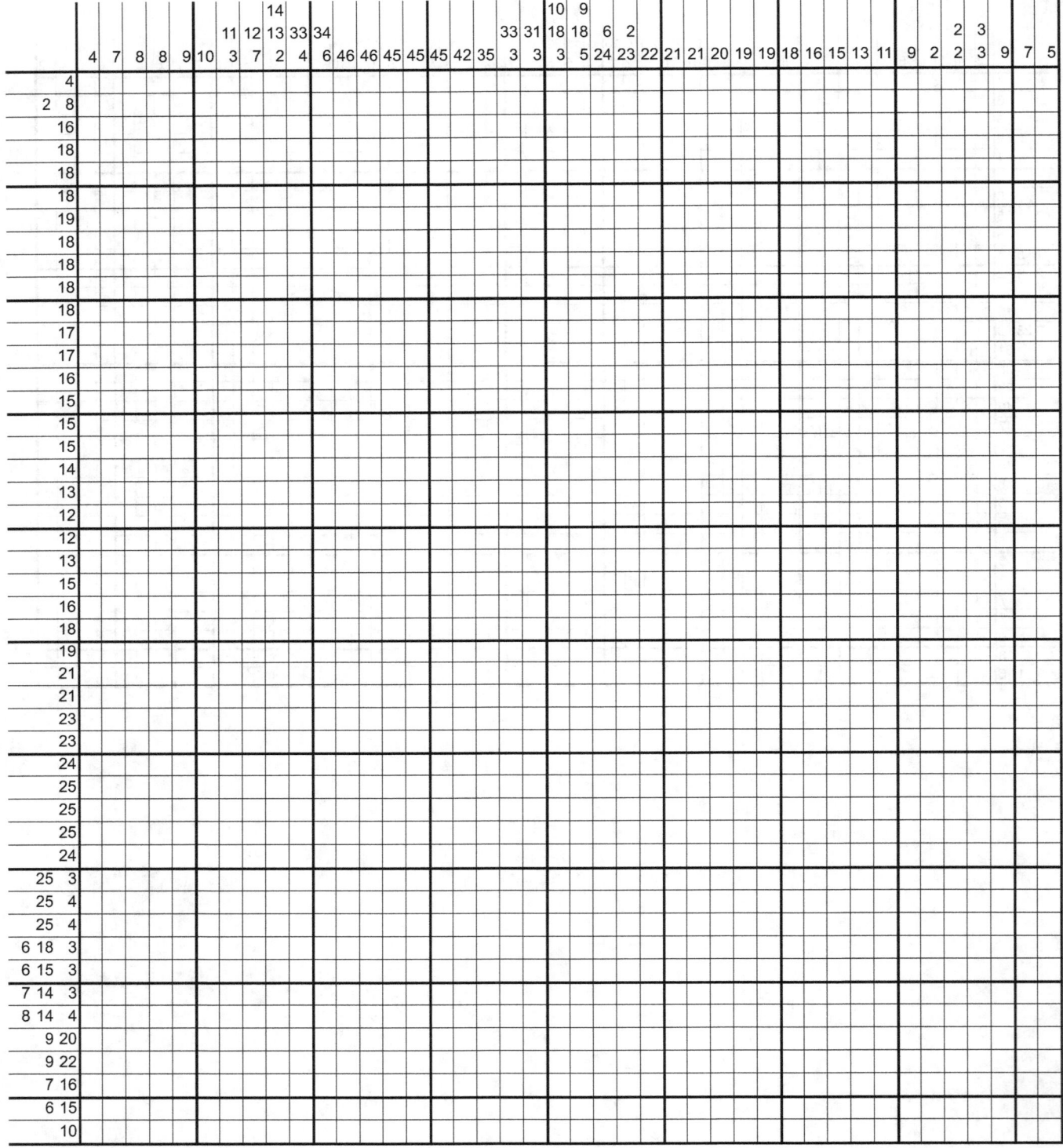

#38

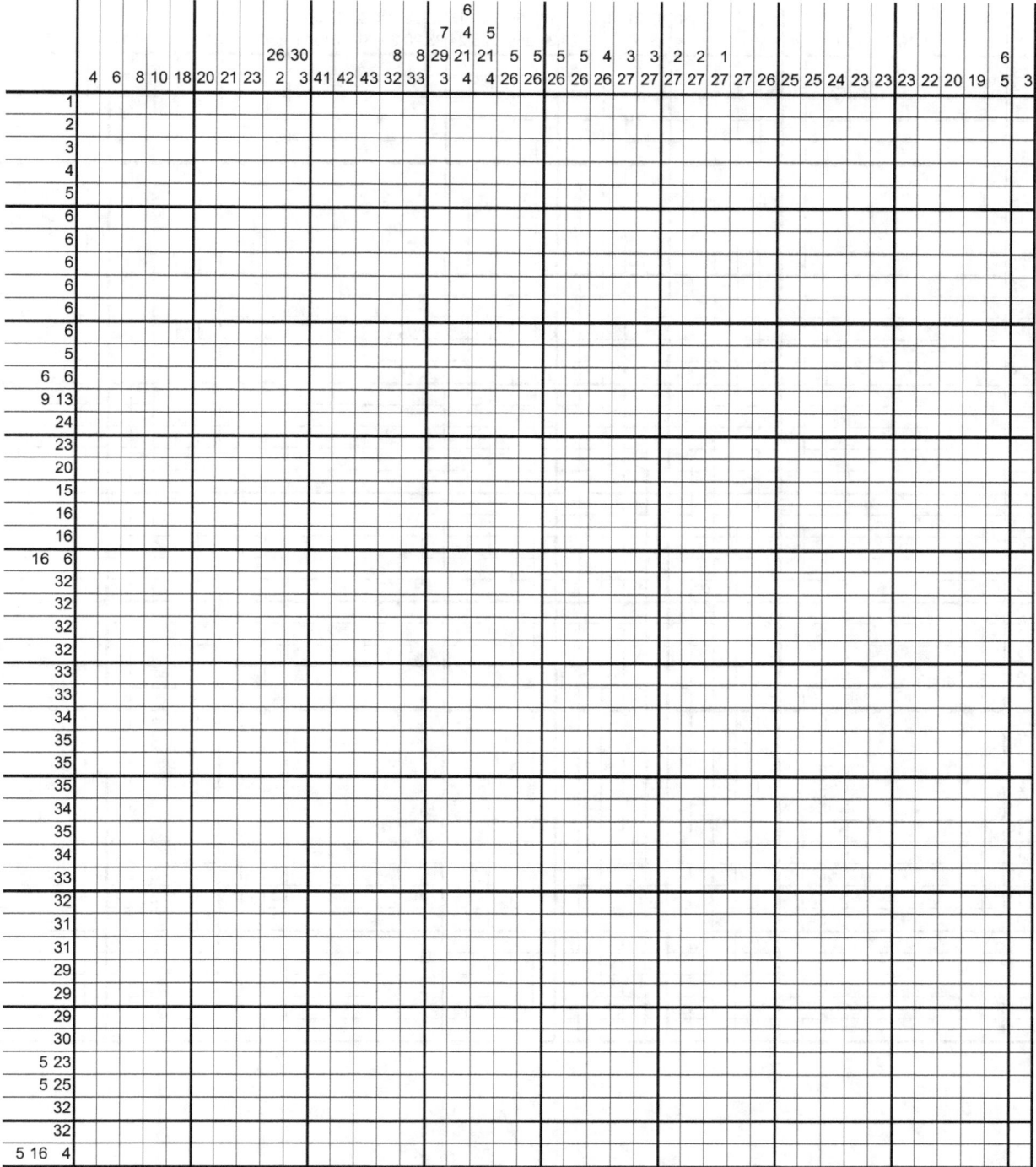

#39

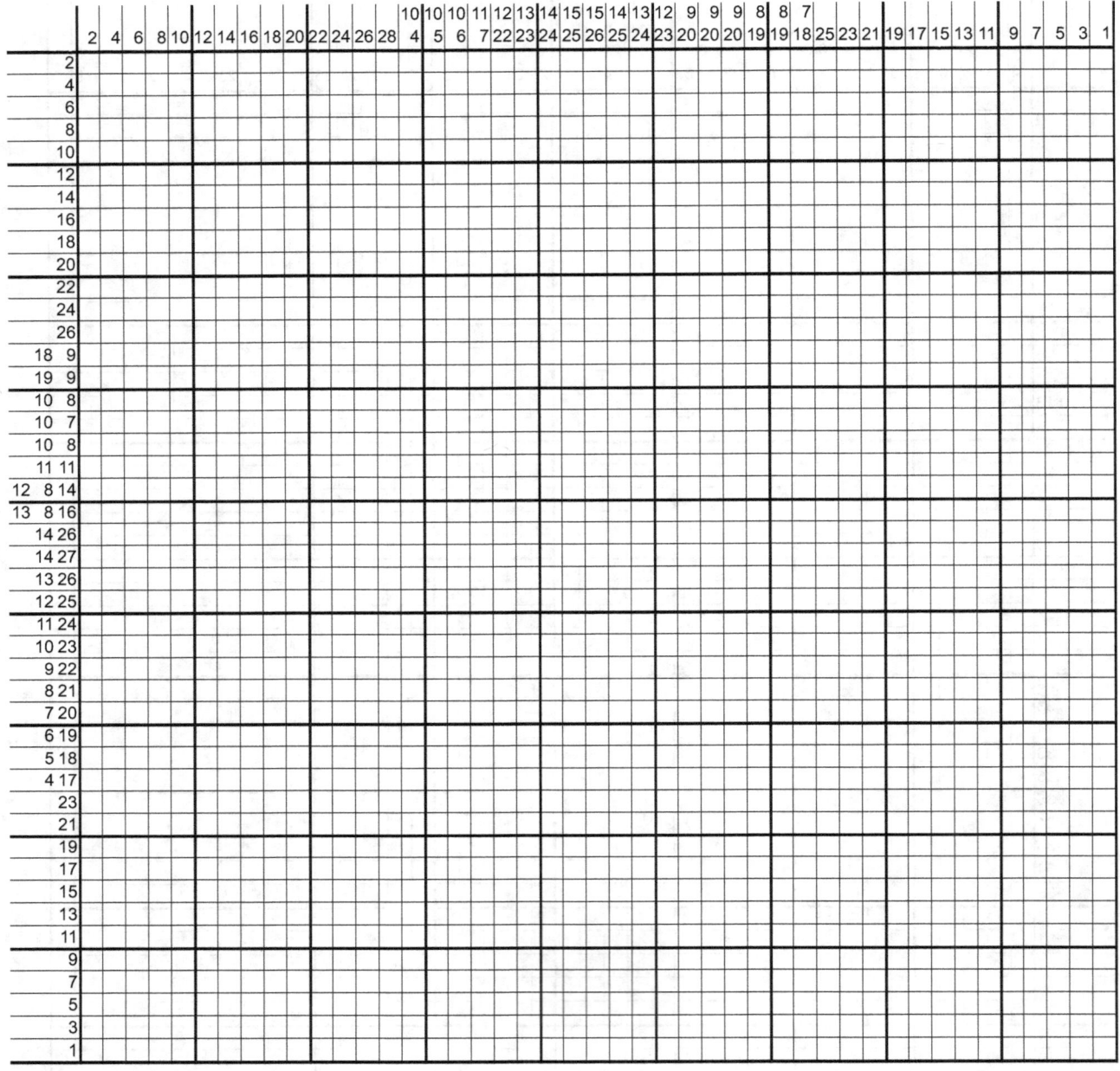

#40

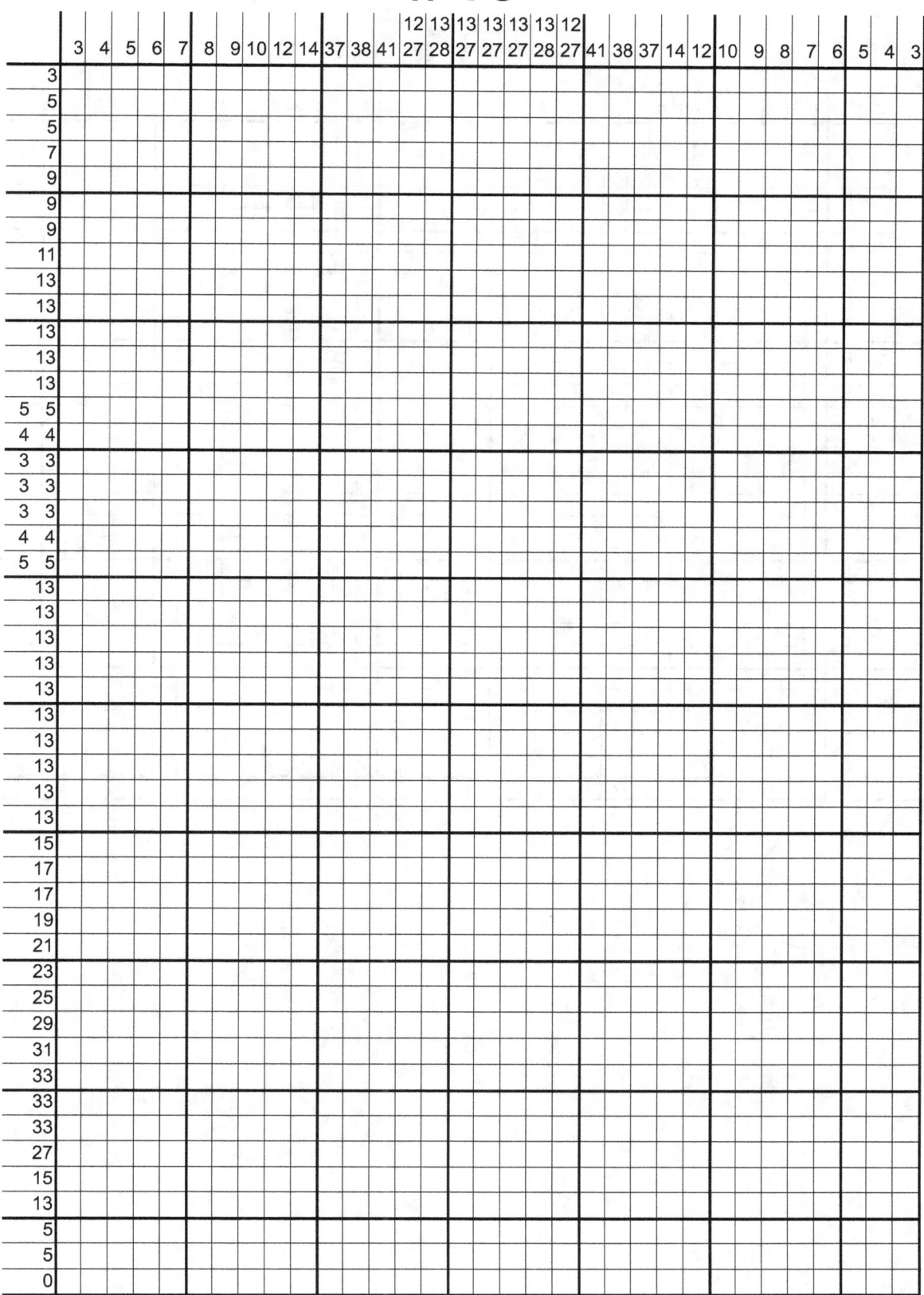

#41

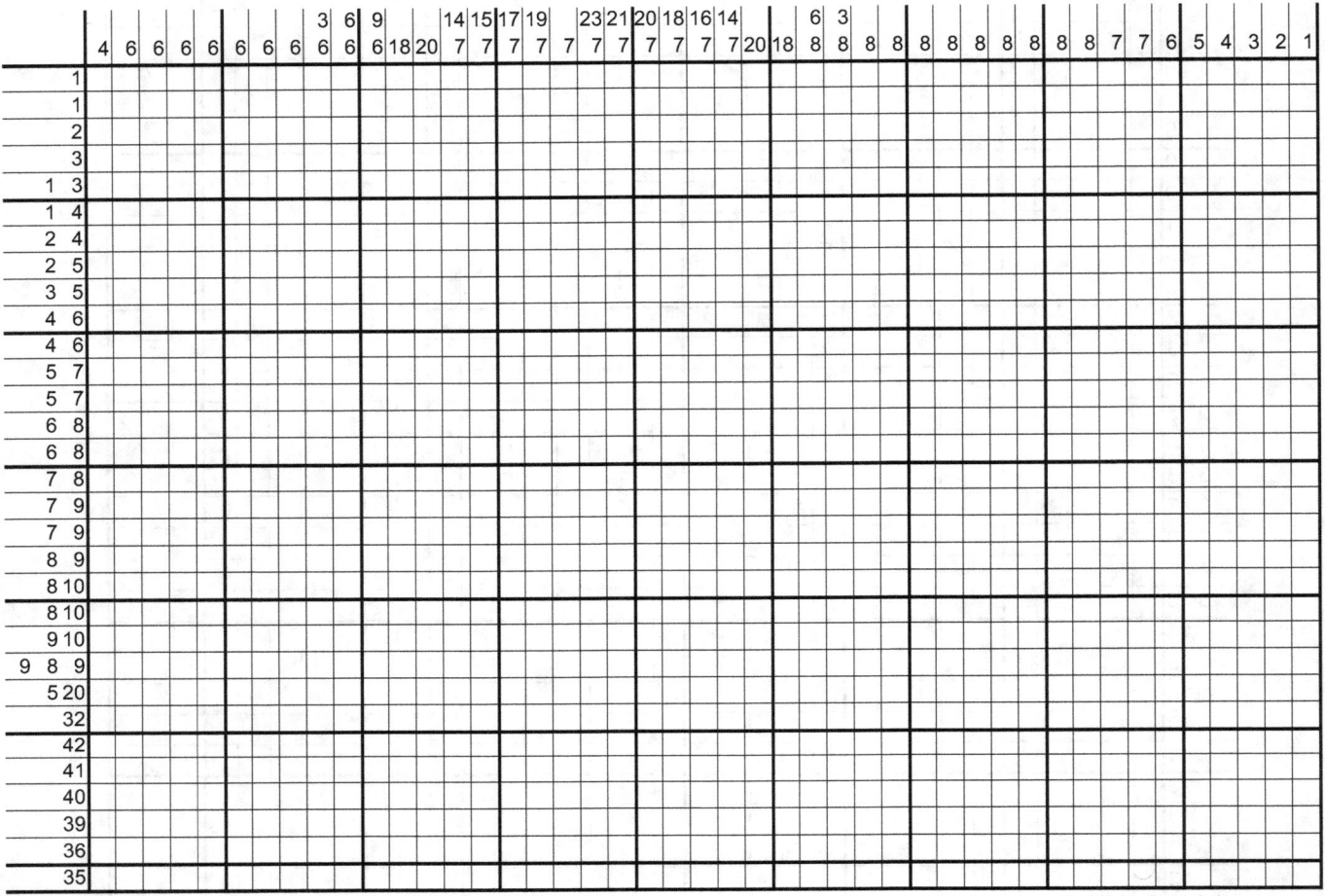

#42

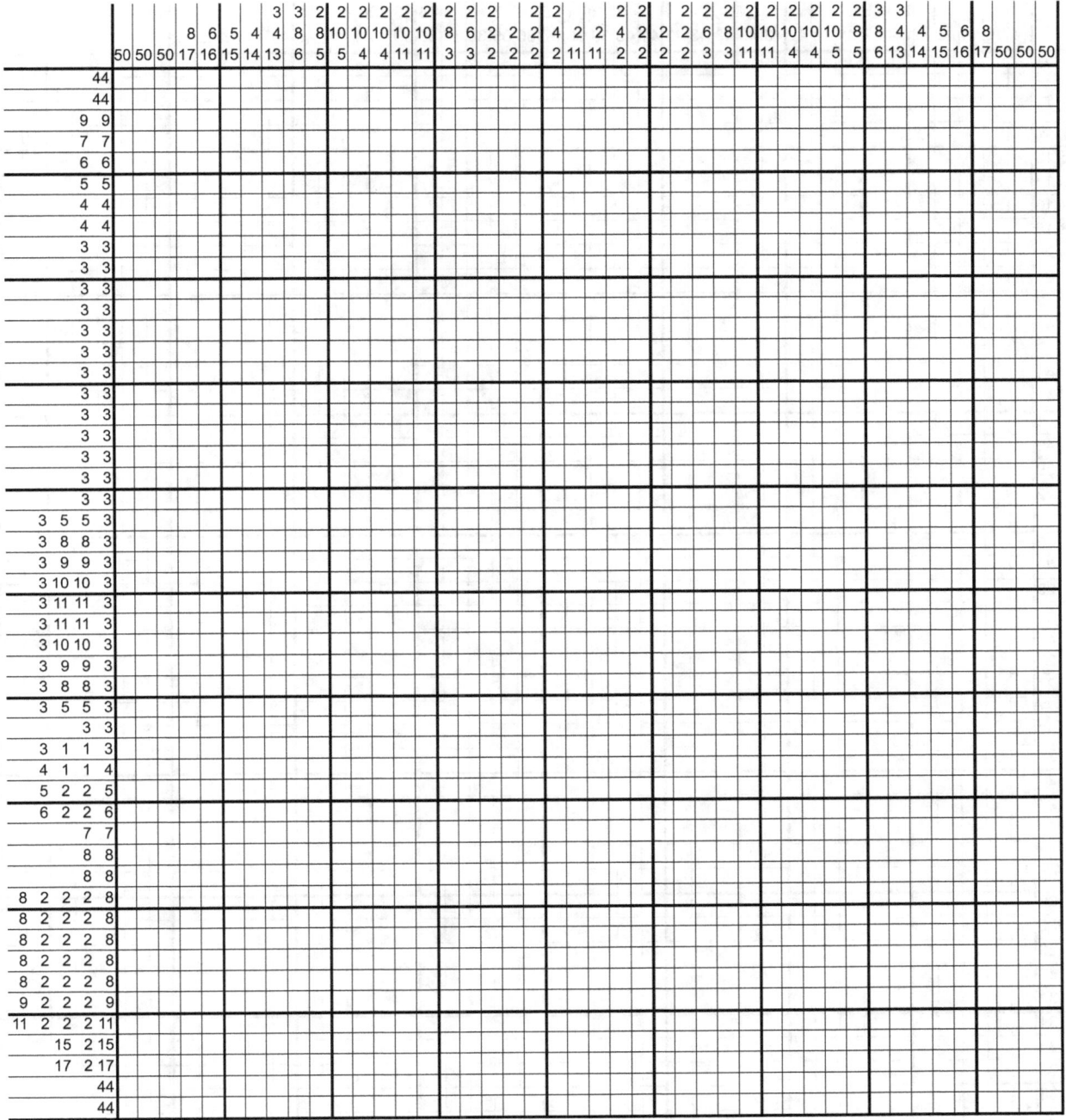

#43

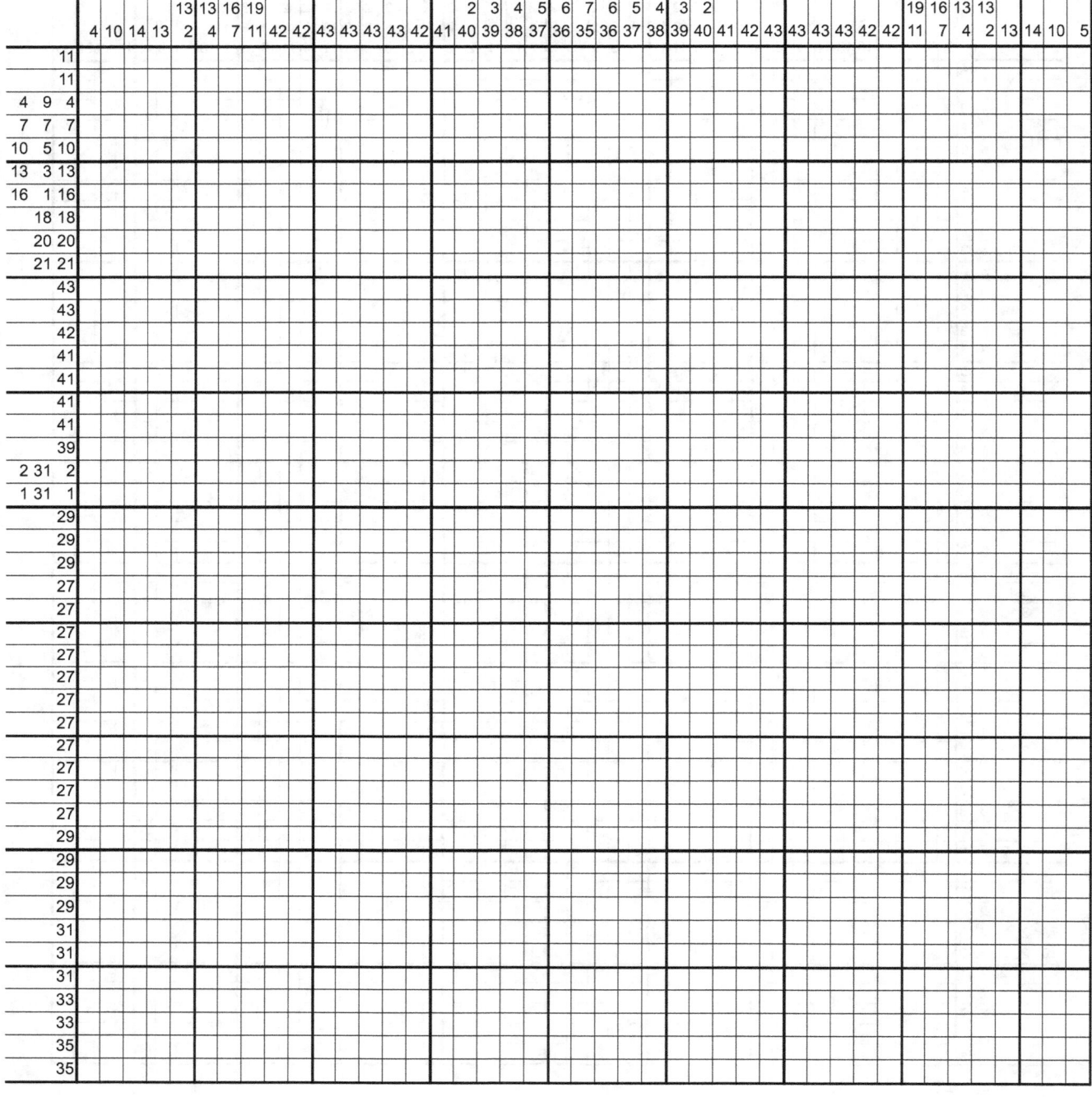

#44

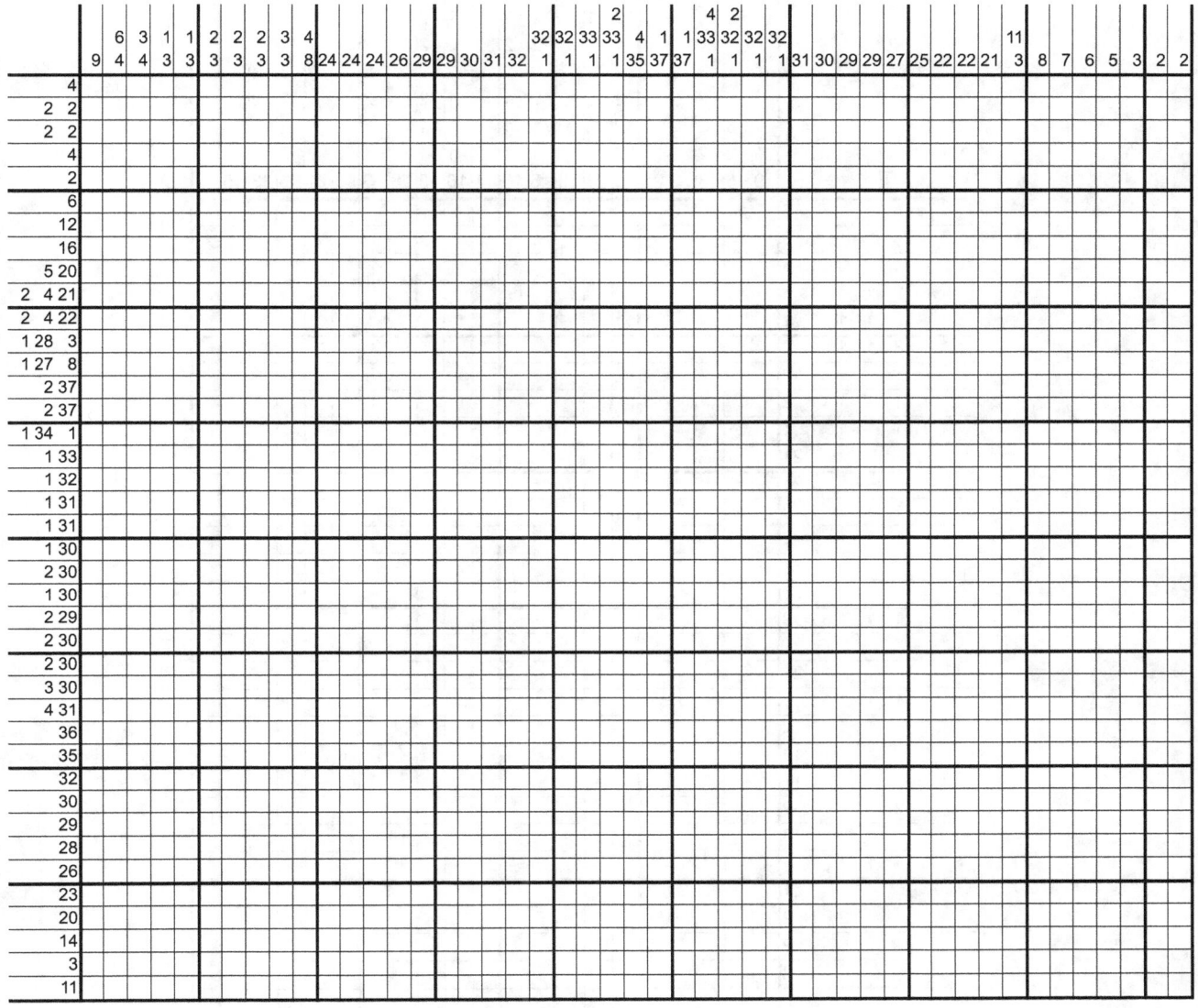

#45

Column clues (left to right, 17 columns):

Col	Clues
1	9
2	5, 2, 5
3	3, 2, 2, 2, 3
4	3, 2, 2, 2, 3
5	21
6	2, 2, 2, 2, 2, 3
7	2, 1, 2, 2, 2, 1, 3
8	28
9	1, 1, 2, 2, 2, 1, 1, 5
10	2, 1, 1, 2, 2, 2, 2, 1, 1, 16
11	1, 28
12	2, 1, 1, 2, 2, 2, 2, 1, 1, 16
13	1, 1, 2, 2, 2, 1, 1, 5
14	28
15	2, 1, 2, 2, 2, 1, 3
16	2, 2, 2, 2, 2, 3
17	21
18	3, 2, 2, 2, 3
19	3, 2, 2, 2, 3
20	15
21	9

Row clues (top to bottom):

1. 3
2. 9
3. 3 1 3
4. 2 1 1 1 2
5. 2 1 1 1 2
6. 3 1 1 1 3
7. 19
8. 2 1 1 1 1 1 2
9. 1 1 1 1 1 1 1
10. 21
11. 21
12. 1 1 1 1 1 1 2
13. 1 1 1 1 1 1 2
14. 21
15. 21
16. 1 1 1 1 1 1 2
17. 21
18. 21
19. 1 1 1 1 1 1 1
20. 2 1 1 1 1 1 2
21. 19
22. 17
23. 2 1 1 1 2
24. 13
25. 1 1 1 1 1
26. 11
27. 2 1 2
28. 9
29. 2 2
30. 2 2
31. 2 2
32. 5
33. 3
34. 3
35. 3
36. 3
37. 3
38. 3
39. 3
40. 3
41. 3
42. 3
43. 3
44. 3
45. 3

#46

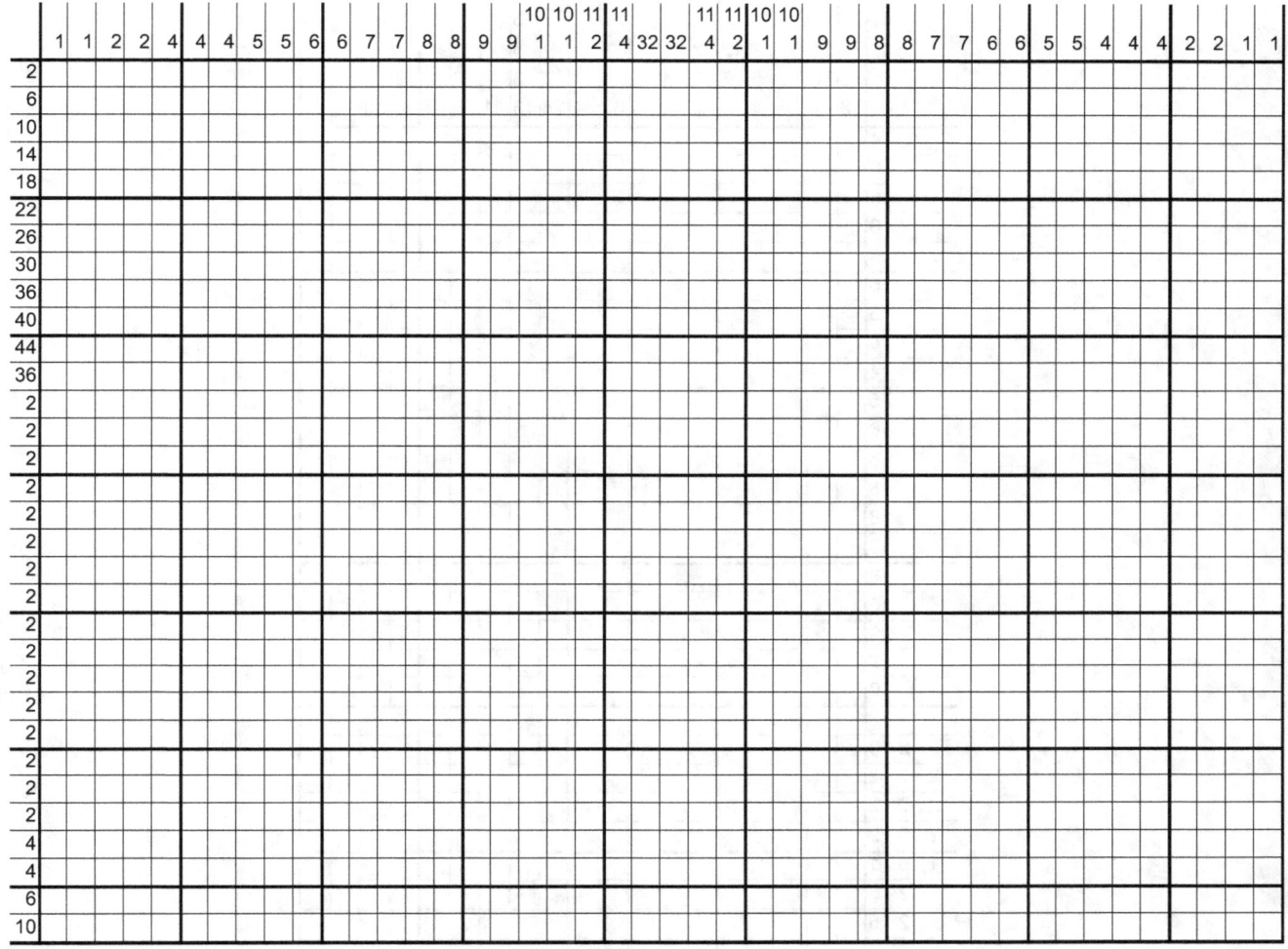

#47

Column clues

					7	12								1 3 5							
	2	8	9		6	7	18	4				7	3	2	2	3	3				
	3	7	9	1	1	1	9	31	34	33	34	28	13	11	8	7	5	3	2	1

Row clues

- 4
- 6
- 5
- 6
- 6
- 5
- 5
- 4
- 4
- 5
- 6
- 6
- 8
- 8
- 9
- 9
- 10
- 9
- 9
- 10
- 12
- 2 10
- 2 7 2
- 2 7 2
- 2 6 2
- 1 6 1
- 2 6
- 2 5
- 6
- 7
- 7
- 7
- 8
- 4 4
- 3 3
- 3 3
- 4 3
- 3 4
- 3 3
- 3 3
- 3 4
- 3 3
- 2 4
- 3 3
- 3 4
- 3 4
- 4 5

#48

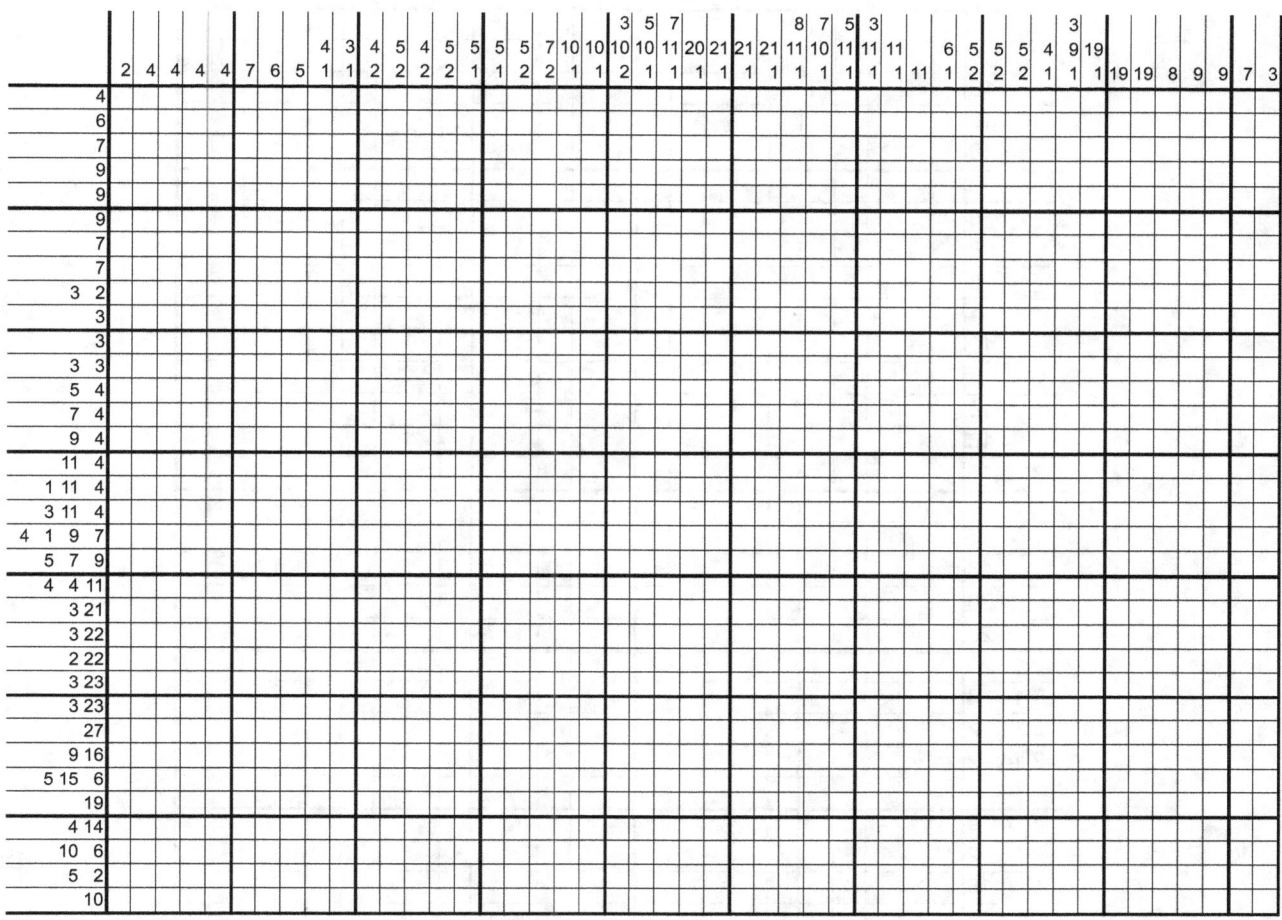

#49

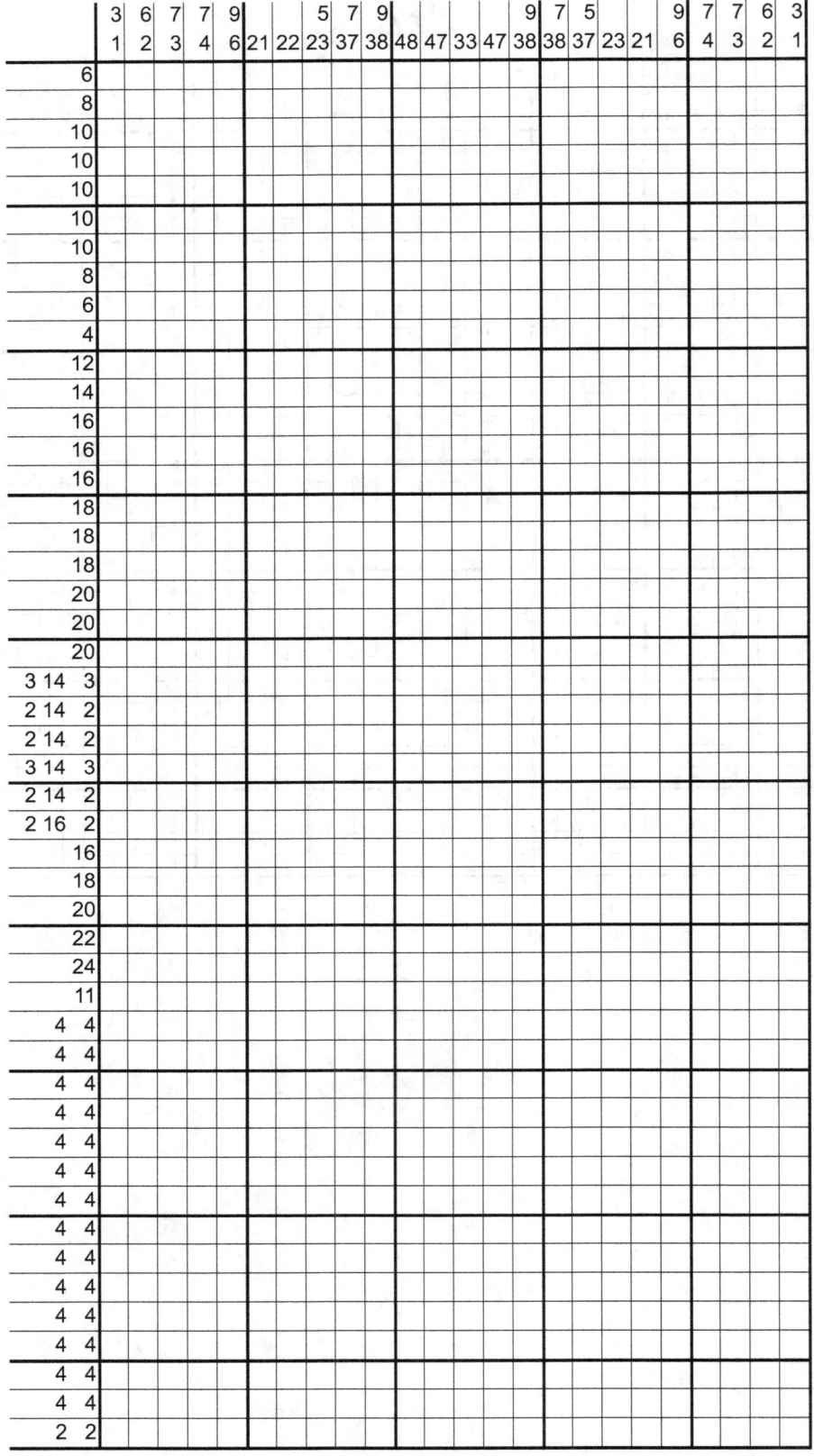

#50

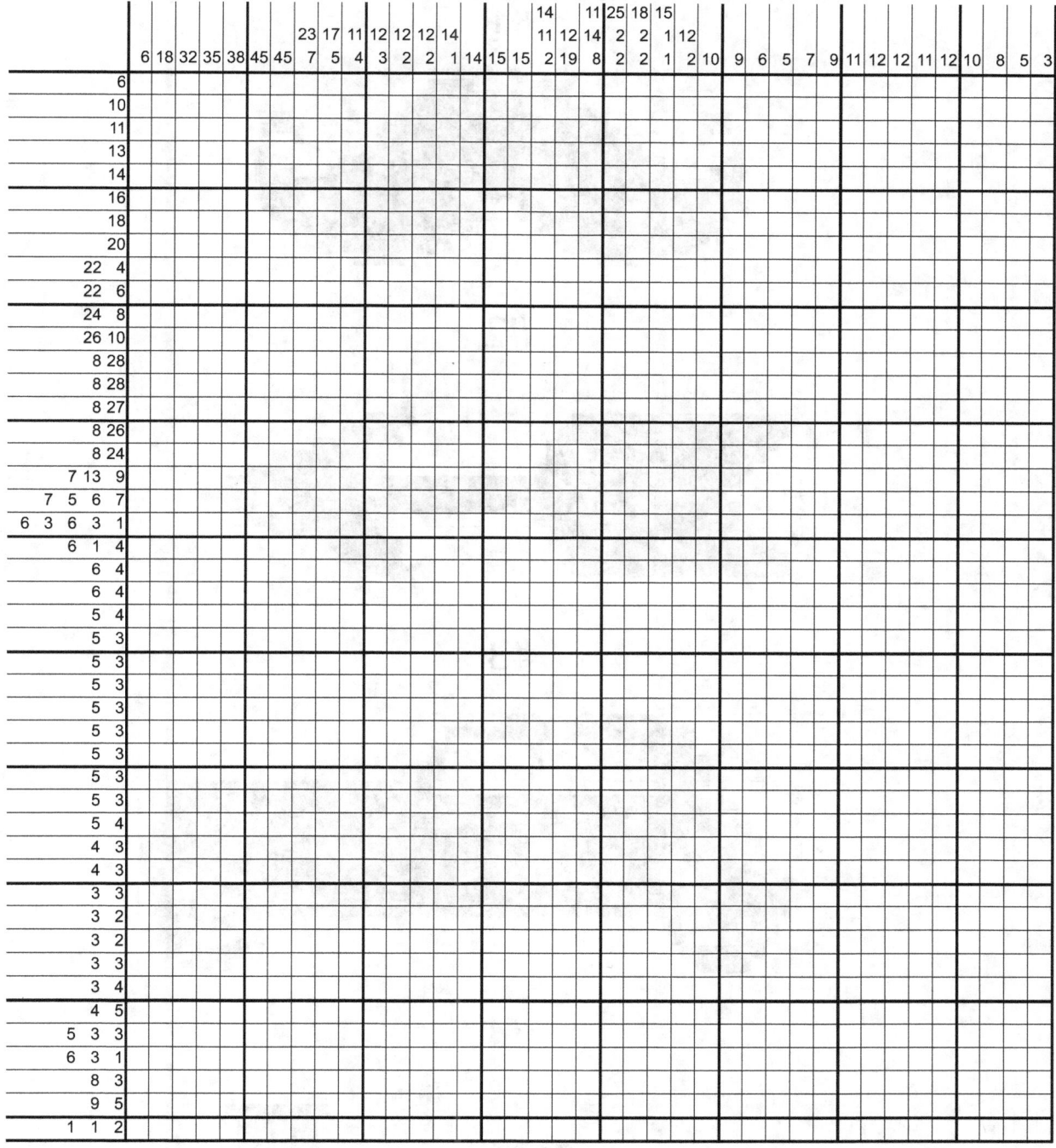

SOLUTIONS

#1

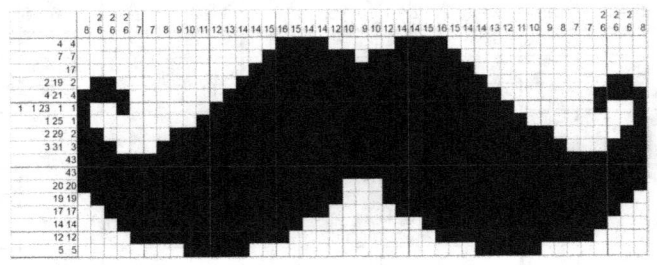

#2

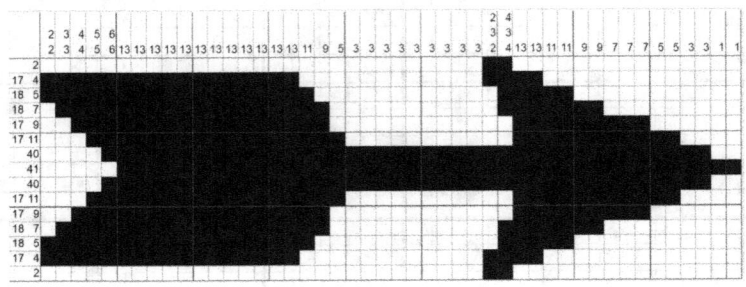

#3

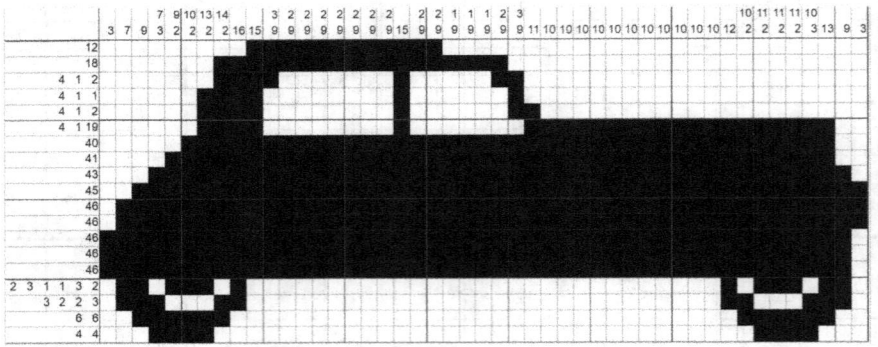

#4

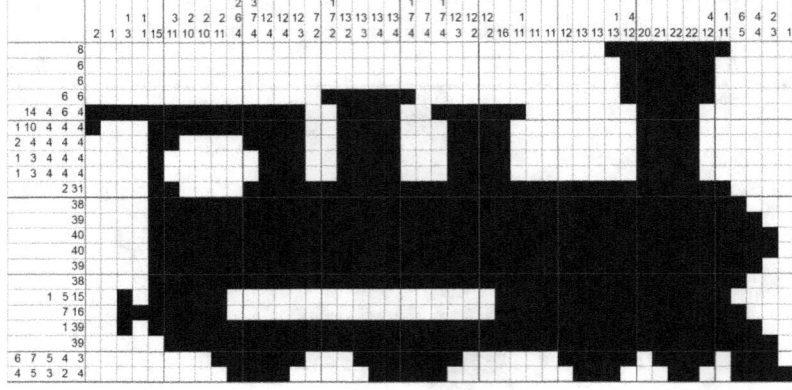

#5 #6 #7

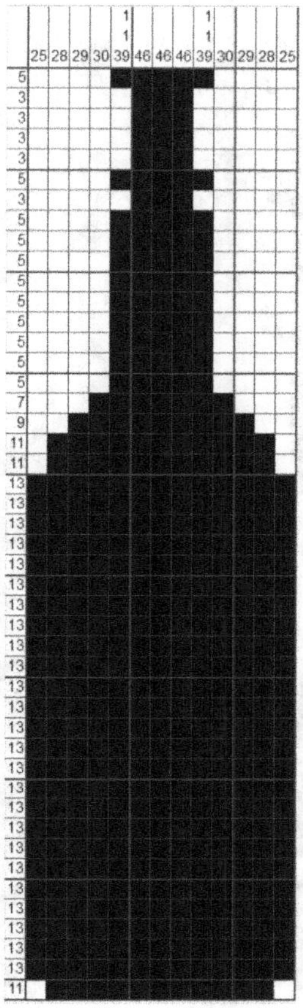

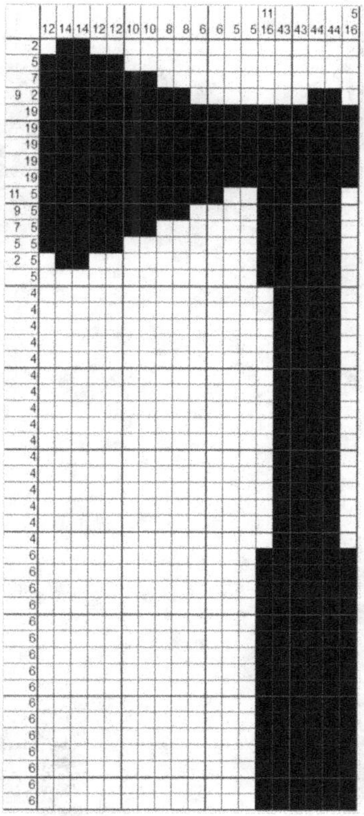

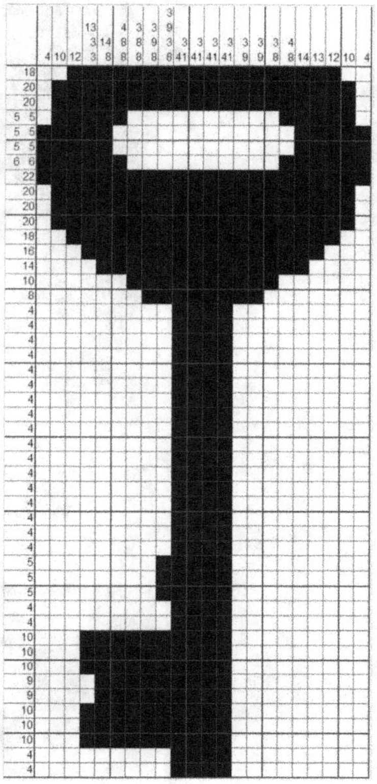

#8 #9

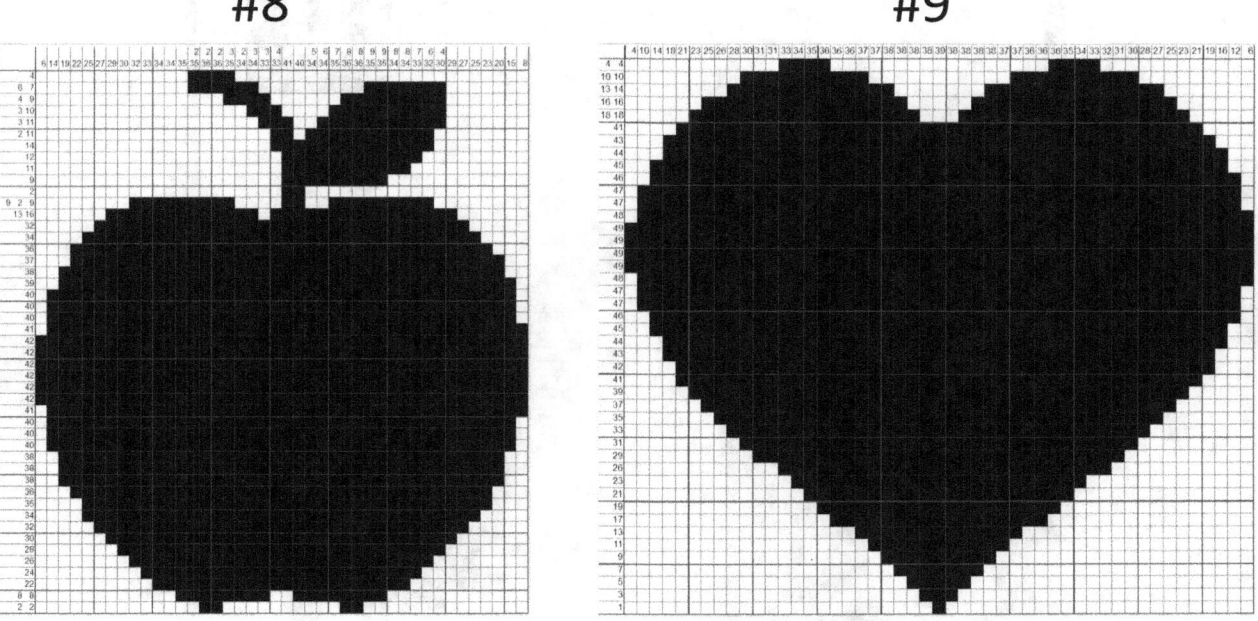

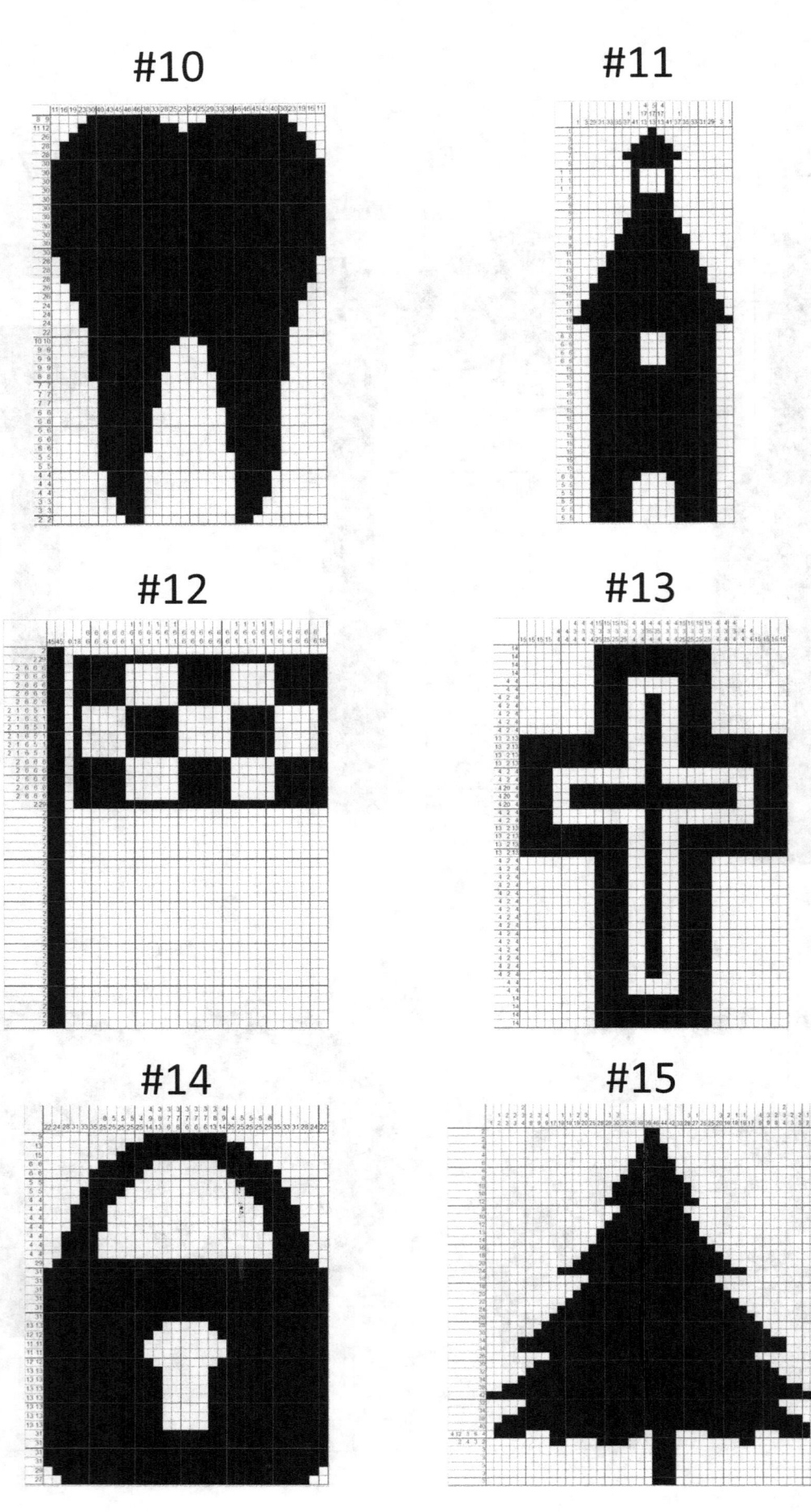

#16
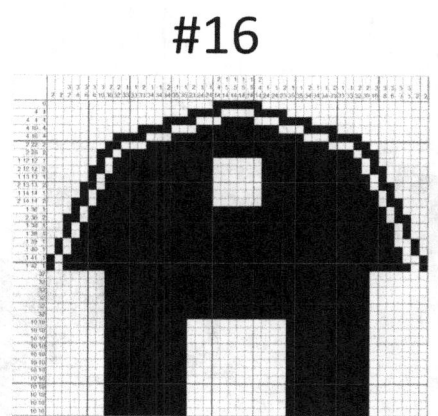

#17
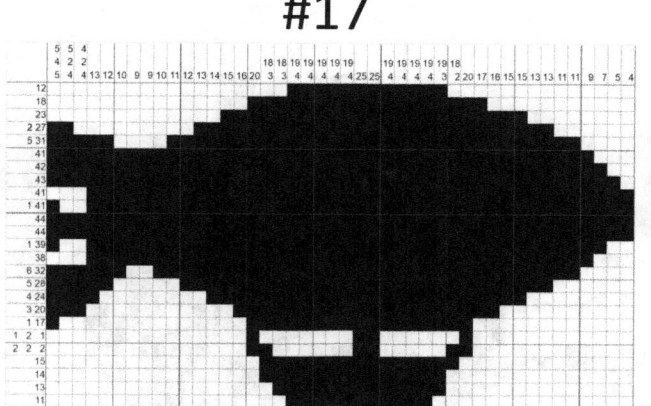

#18
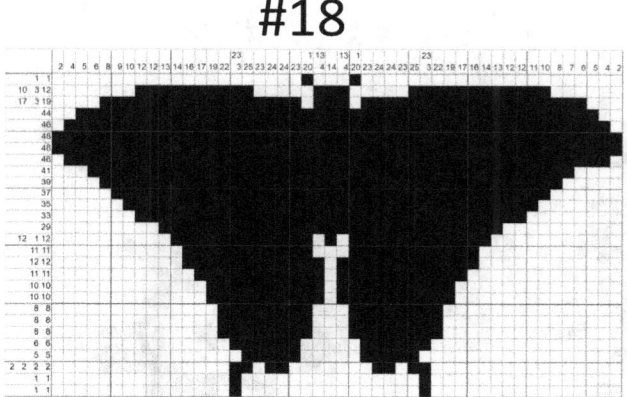

#19
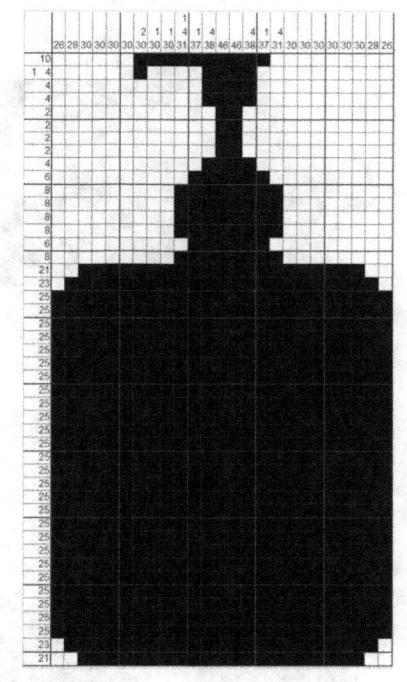

#20
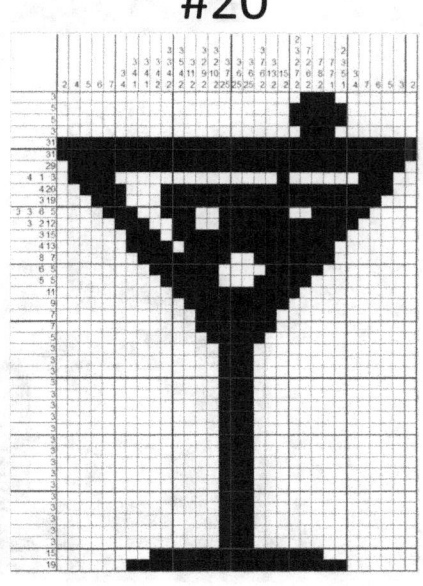

#21
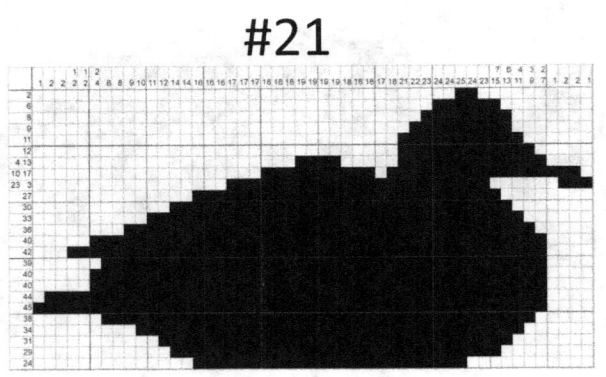

#22
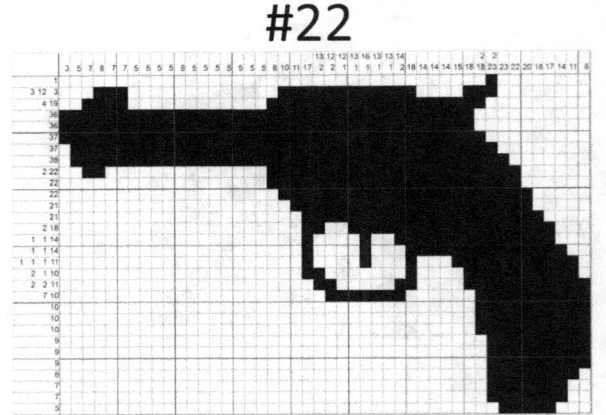

#23
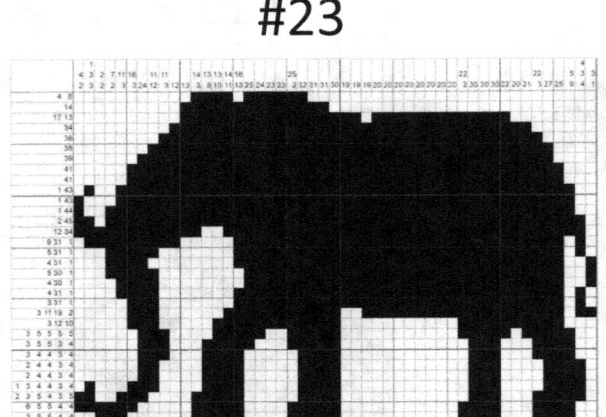

#24
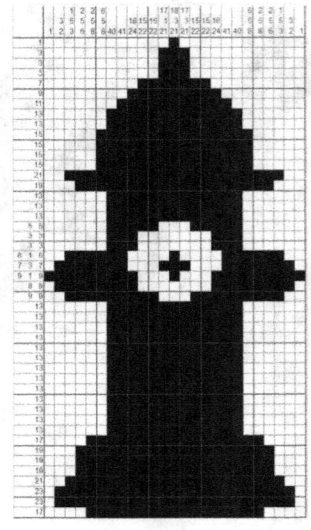

#25
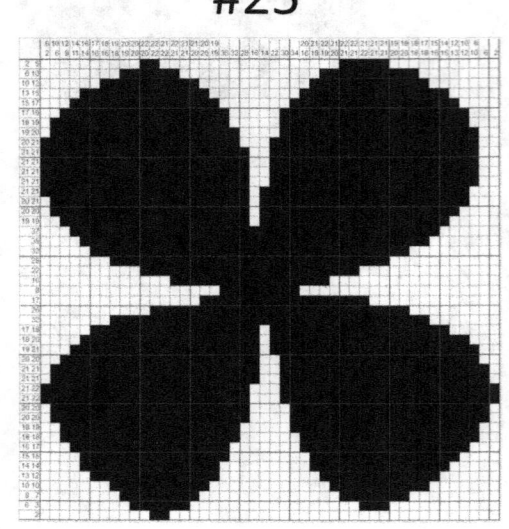

#26
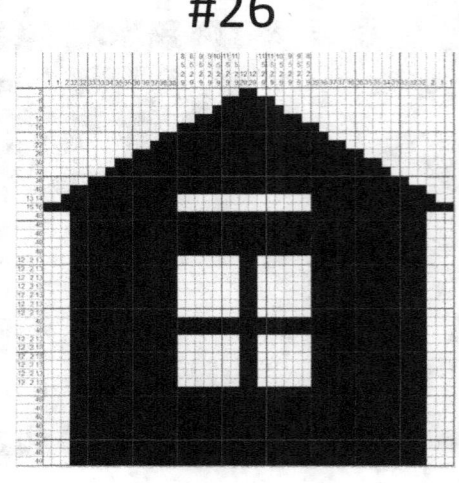

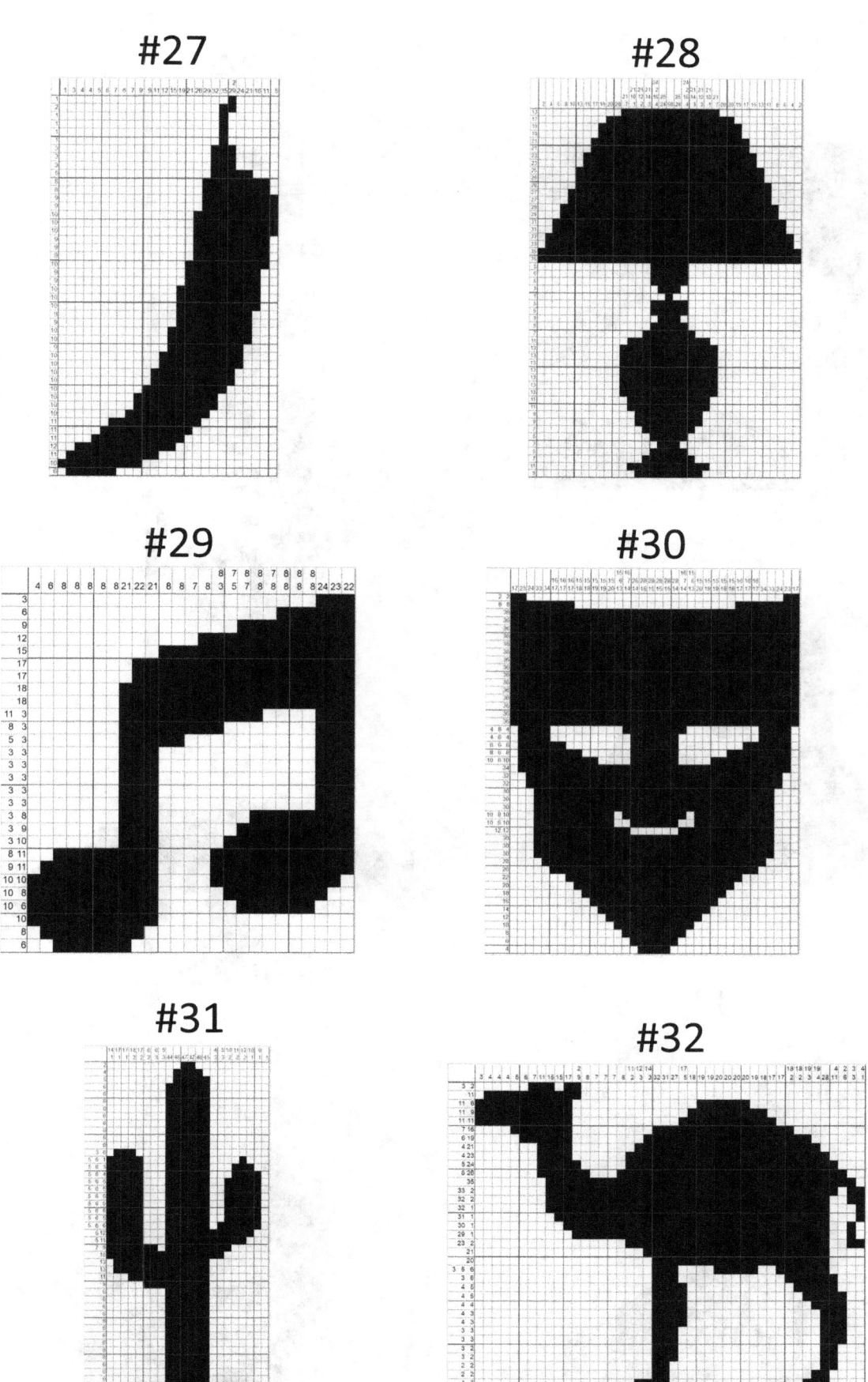

#33
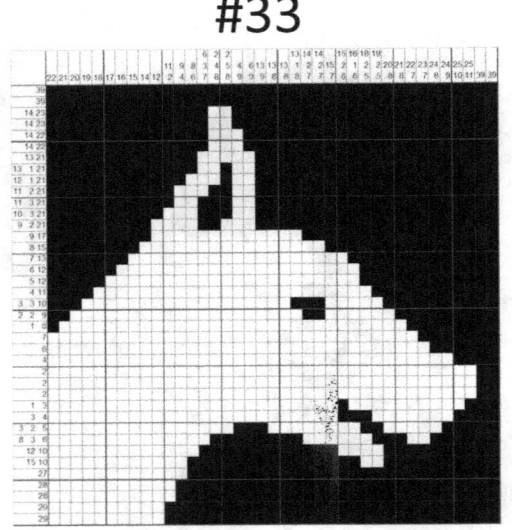

#34
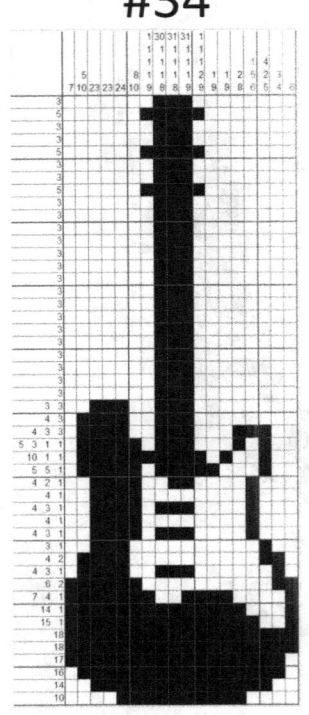

#35
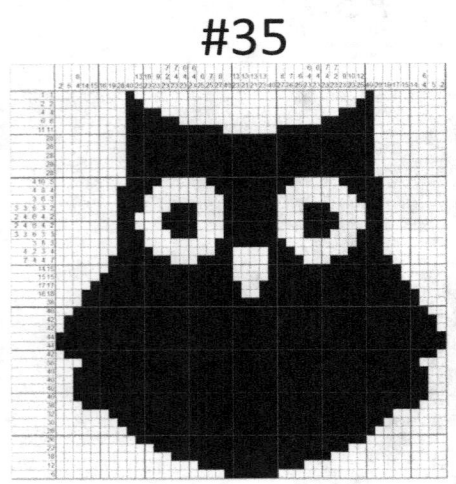

#36
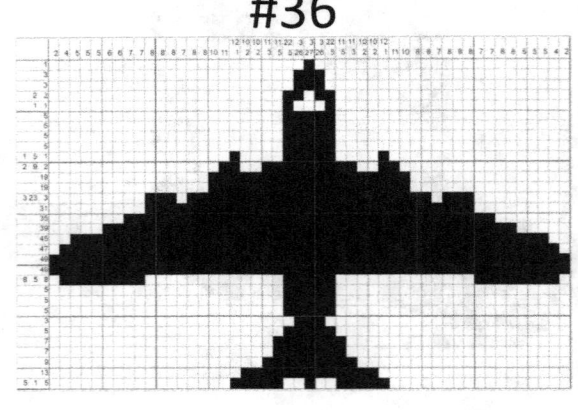

#37
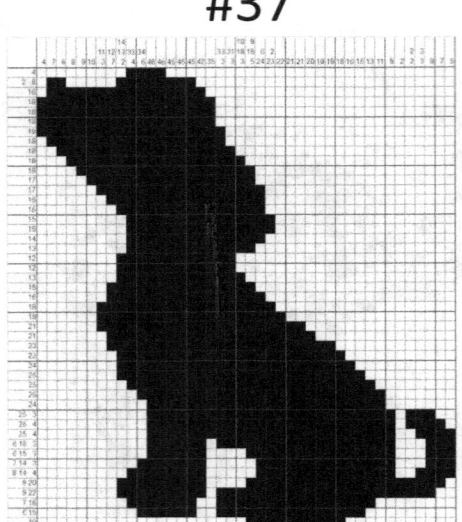

#38
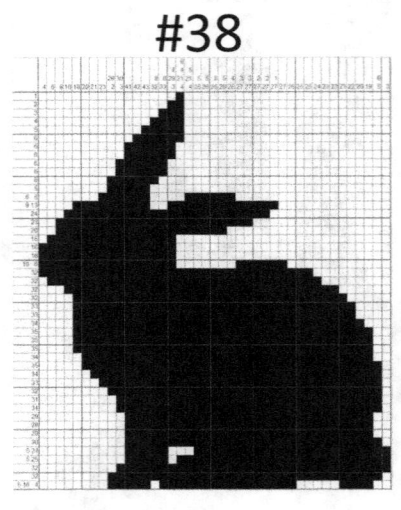

#39
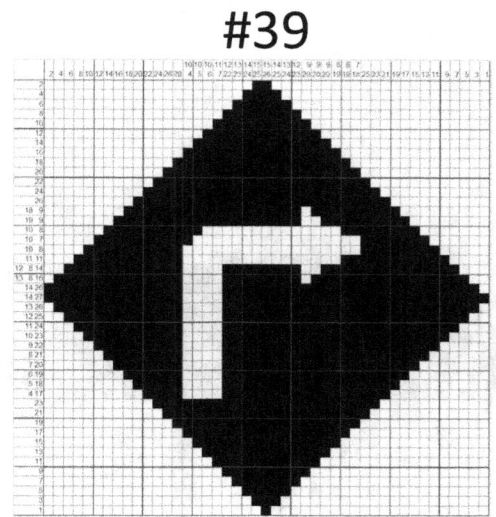

#40
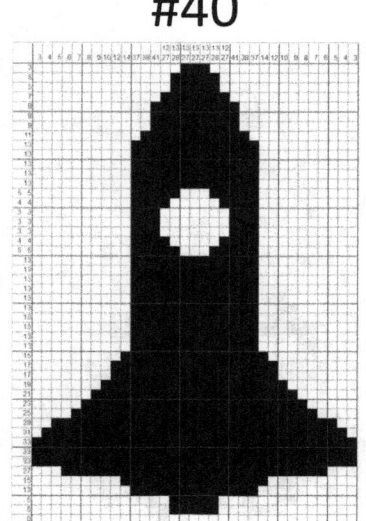

#41
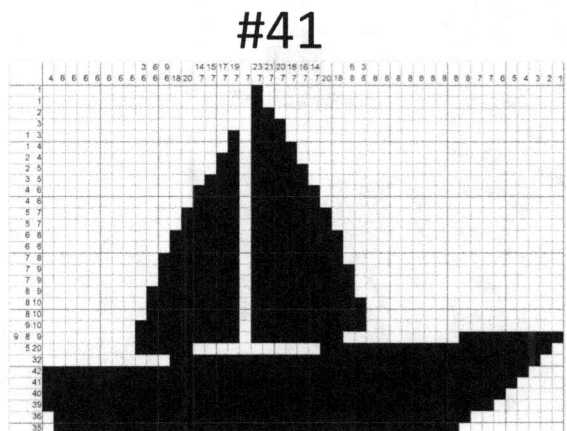

#42
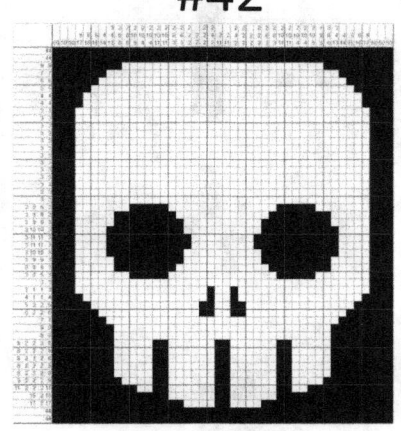

#43
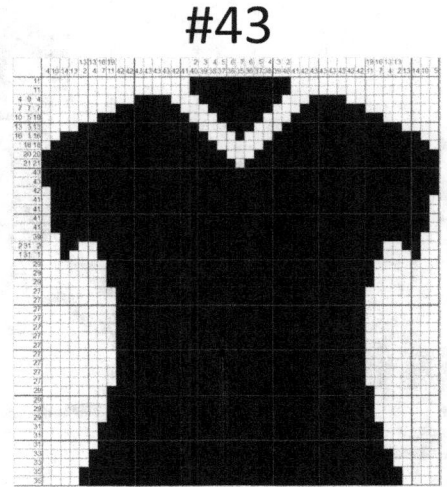

#44
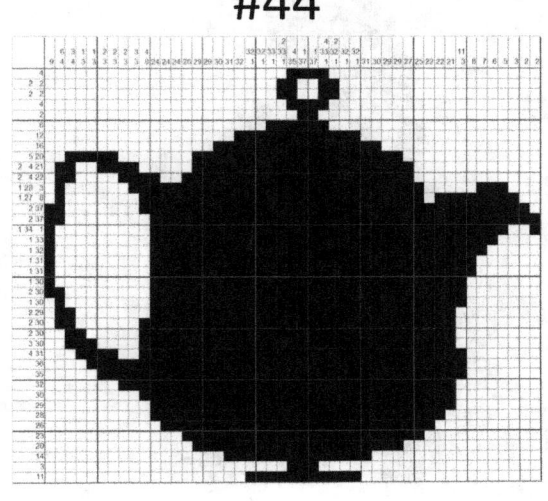

#45

#46
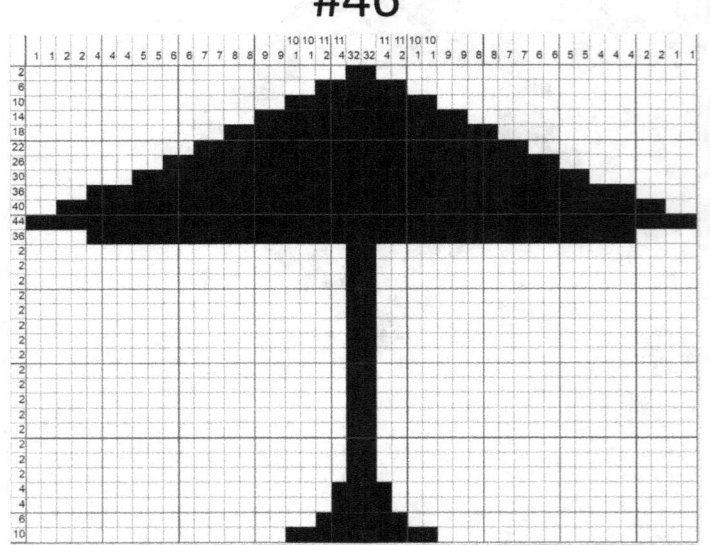

#47
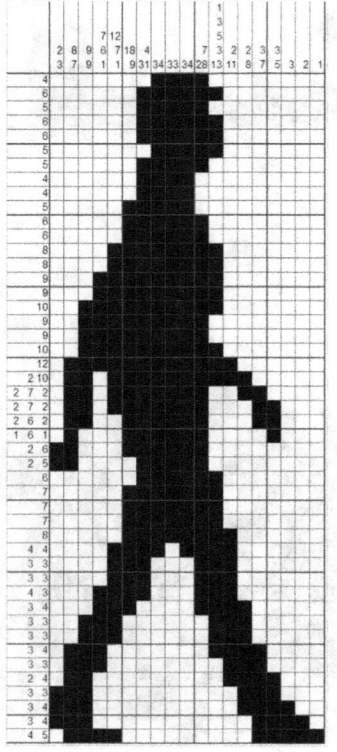

#48
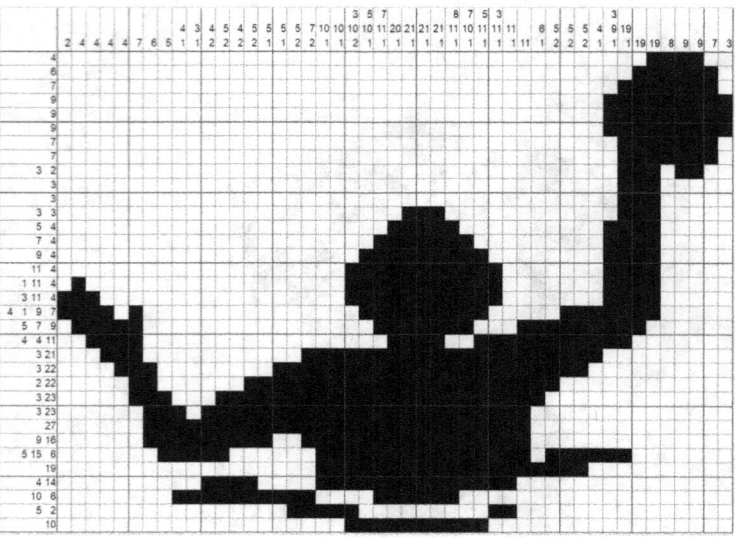

#49

#50

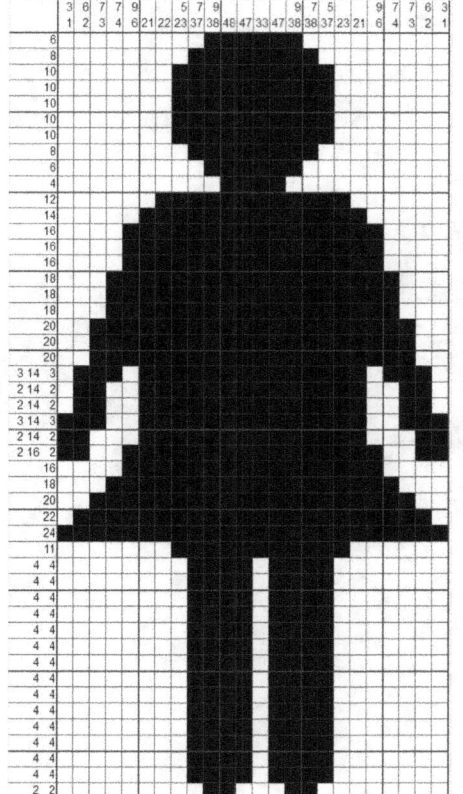
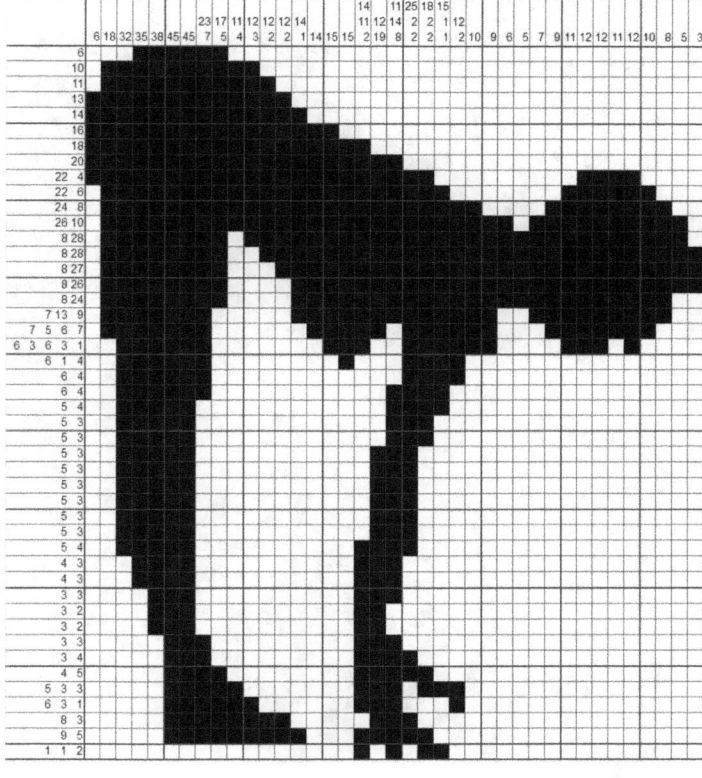

www.ingramcontent.com/pod-product-compliance
Lightning Source LLC
Chambersburg PA
CBHW080623220526
45466CB00010B/3443